Synthese Library

Studies in Epistemology, Logic, Methodology, and Philosophy of Science

Volume 499

The aim of *Synthese Library* is to provide a forum for the best current work in the methodology and philosophy of science and in epistemology, all broadly understood. A wide variety of different approaches have traditionally been represented in the Library, and every effort is made to maintain this variety, not for its own sake, but because we believe that there are many fruitful and illuminating approaches to the philosophy of science and related disciplines.

Special attention is paid to methodological studies which illustrate the interplay of empirical and philosophical viewpoints and to contributions to the formal (logical, set-theoretical, mathematical, information-theoretical, decision-theoretical, etc.) methodology of empirical sciences. Likewise, the applications of logical methods to epistemology as well as philosophically and methodologically relevant studies in logic are strongly encouraged. The emphasis on logic will be tempered by interest in the psychological, historical, and sociological aspects of science. In addition to monographs *Synthese Library* publishes thematically unified anthologies and edited volumes with a well-defined topical focus inside the aim and scope of the book series. The contributions in the volumes are expected to be focused and structurally organized in accordance with the central theme(s), and should be tied together by an extensive editorial introduction or set of introductions if the volume is divided into parts. An extensive bibliography and index are mandatory.

Charles H. Pence

Integrative Promise

Explanatory Virtues in Biology

Springer

Charles H. Pence
Institut supérieur de philosophie
Université catholique de Louvain
Louvain-la-Neuve, Belgium

ISSN 0166-6991 ISSN 2542-8292 (electronic)
Synthese Library
ISBN 978-3-031-88470-2 ISBN 978-3-031-88471-9 (eBook)
https://doi.org/10.1007/978-3-031-88471-9

This Springer imprint is published by the registered company Springer Nature Switzerland AG
The registered company address is: Gewerbestrasse 11, 6330 Cham, Switzerland

For Julia, so she can keep telling the book joke

Preface

The book that you're now holding is a rather strange creature. At least some of that strangeness deserves explication and apology (both in the classic sense of "defense" and the colloquial sense of "excuse") before you proceed any further. Let me start with a provocation—not to you, but to me. A friend, whom I won't name lest he think that I mean this as a criticism, asked a question that I imagine a number of readers might themselves have when picking up this book: do we really need another book on explanations in biology? To this, I respond that—despite the title—I will not propose here a novel theory of explanatory virtues in biology, nor a thorough description of all such virtues, nor an argument against other theories of those virtues already elaborated in the literature.

My purpose is different. What I want to do here is offer a new tool for the analysis of explanations in the life sciences, drawn from one central insight: that there is a repeated pattern of use of a set of explanatory virtues that are each already relatively well known—a pattern that is instantiated equally well across centuries of the history of biology and across disciplines as diverse as molecular biology, ecology, and infectious disease research. I argue here that explanations in the life sciences are both defended, after the fact, and promoted, in advance, by appeals to a collection of explanatory virtues that I will baptize *integrative promise*.

First, a brief synopsis of the content of that pattern. Biologists, I claim, often advance explanations because they will *integrate* our knowledge of the biological world, whether horizontally (across the tree of life) or vertically (across something like "levels" of composition, organization, or scale). Those explanations regularly take advantage of *opportunistic* connections to existing biological, experimental, or theoretical knowledge, but they do so while recognizing the constraint that—since integrative knowledge is extremely hard to obtain—they must put their proposed explanations to *stringent* tests. And they do not go to all this work idly; such integrative explanations regularly make reference to the social or ethical consequences of obtaining this integrative knowledge—that is, they wear their *non-epistemic value* commitments on their sleeve. Explanations that are defended for exemplifying this whole set of interlinked explanatory virtues are, in my coinage, promoted for their *integrative promise*.

While I'll canvass a number of cases of integrative promise and its components in depth later on, let me sketch, with an intellectually negligent level of brevity, an example of the kind of thing that I have in mind. Take the explanations for the evolution of morphological traits proposed by evolutionary developmental biology. These explanations are presumed to integrate not only resources at multiple "levels" or "scales" of the living world—from the biophysical influences of shear forces, through biochemical gradients and low-level gene regulatory processes, on to the body plans of organisms and subsequent effects on ecosystem structure and major evolutionary events like the Cambrian explosion—but also, largely via evolutionary conservation of key developmental ingredients, to integrate our knowledge of those phenomena across the tree of life.

These explanations are opportunistic, in that they draw upon enormous existing bodies of theory and practice both in ecology and evolutionary biology and in developmental and molecular biology, but they're also cognizant of the need to subject their theoretical proposals to particularly stringent tests, to avoid the charge of spinning just-so stories. They are also frequently directed at the support of non-epistemic values. The abandonment of gene-first and gene-centric perspectives of evolution and development are, I think quite rightly, promoted as antidotes to reductive pictures driven by genetic determinism and "genes for" everything from human sexuality to complex character traits like intelligence.

This, in short, is what I take integrative promise to exemplify—the pursuit of this collection of interlinked and interacting virtues in explanations across the life sciences. But the case is, of course, not quite so simple. In the first part of the book, I'll start by doing some ground-clearing, describing in Chap. 1 the ways in which my work here is related to some other nearby efforts in general philosophy of science (especially other work on theoretical virtues and on pursuit-worthiness).

Then, in Chaps. 2–4, I'll explore the various components of integrative promise that I've identified above. While a colloquial sense of integration as I've used it here is intuitive enough, philosophers of science will be well aware that depictions of discrete "levels" of organization or composition, which seem to underlie that intuition, have been subject to extensive and decisive philosophical criticism. I'll therefore generalize that intuitive view to a broader notion that I'll call *inter-domain* integration, a concept flexible enough to capture the intuitive cases while not, I hope, falling prey to the now-evident counterexamples.

I'll then present my understandings of opportunism, stringency, and the invocation of non-epistemic values (which I'll often describe—for reasons that will become apparent from the examples that I consider—as an appeal to solutions to "grand challenges"). Finally, in Chap. 5, the first part of the book will conclude with a detailed case study that brings all these components together, on an integrative-promise analysis of fears about the place of "traditional" natural history in an increasingly molecular- and theory-driven biological establishment.

The second part of the book, in Chaps. 6–8, turns outward to look at ways in which using the framework that I've developed can offer fruitful analysis of other, broader questions in contemporary biology: the nature of "big data" or "data-driven" biology; the use and development of model organisms; and the role played in the life sciences by digital analyses of the scientific literature. In the conclusion, I'll deploy integrative promise to look toward the future, making some concrete proposals for how this tool might encourage reforms in biological practice.

I have endeavored to keep as much of the book as possible readable for as many audiences as possible. Philosophical technicalities are largely sequestered in Chaps. 1 and 2. If the reader (especially the reader trained in biology) is uninterested in these details, and is willing to rest content with a colloquial sense of horizontal and vertical integration, then they could start the story with Chap. 3. The chapters in the second half are somewhat independent from each other, showing ways in which using the lens of integrative promise can shed light on other kinds of traditional and not-so-traditional problems in the philosophy and practice of the life sciences. They can be consumed in any order, following the reader's interests.

As I'll note again in the conclusion, this book is largely free of polemic; there is no existing debate over the nature and role of integrative promise (or some similar concept) to which I'm providing an opinionated contribution. I'm rather offering this book in the spirit of something like conceptual engineering in the philosophy of biology. It's my hope that having the framework of integrative promise at hand can allow us to helpfully intervene in some of the most interesting areas of contemporary work in the life sciences.

One more clarification deserves to be raised at the outset. The subtitle of this book is "Explanatory Virtues in Biology," and I've somewhat interchangeably used "biology" and "the life sciences" to describe the potential scope of integrative promise throughout. I don't have any particularly strict demarcation between "life sciences" and "non-life sciences" in mind. Biology is the domain that I know best, and it's the one that inspired me to consider this pattern in the use of explanatory virtues. As I'll make very clear in what follows, I by no means intend to argue that *every* explanation in biology is prized for its integrative promise—far from it. The precise boundaries of the life sciences are thus unimportant for me here. I do not know whether or not integrative promise will seem to be an interesting tool for philosophers considering other branches of the sciences. The ways in which I will present vertical and horizontal integration in Chap. 2 are quite specific to biology, for instance, and would have to be reinterpreted in other fields. But I lack the familiarity with physics, chemistry, or the social and behavioral sciences required to do that evaluation myself.

I have been lucky enough to work on a number of parts of this project (sometimes in an embarrassingly early state) with a variety of talk audiences. The entire concept derives from my participation at the European Molecular Biology Laboratory (EMBL) Graduate Conference in Heidelberg in 2019; my sincere thanks to the organizers and everyone with whom I had the chance to chat over the days of the meeting. The deep commitment to integrative promise that I saw there was, and remains, inspiring.

The considerations around interdisciplinarity in Chaps. 2 and 9 were refined in two talks; thanks to an audience at the Third DR2 Conference at Sapiena Università di Roma, especially Stefan Heßbrüggen-Walter, Guido Bonino, and Paolo Tripodi; and at the Max Planck Institute for the History of Science's Final Theory Program Seminar, especially Pablo Ruiz de Olano and Alexander Blum. My reflections on science and values in Chap. 9 owe much to chats with biology master's degree students here at UCLouvain, in the course LBOE2125, "Biodiversity and the Human."

Chapter 7 contains material from Evans, Nicholas G. and Pence, Charles H., 2024, "Gain of Function Research and Model Organisms in Biology," *Journal of Medical Ethics* 50(3): 201–206, https://doi.org/10.1136/jme-2022-108853, with agreement of the publisher. My many thanks to Nick for being a fantastic collaborator and friend. For this chapter in particular, I'd like to thank Rachel Ankeny for discussions on the project; she is also exploring this case study, and we eagerly await her results. Thanks as well to an audience at the conference "Epancopi—Enjeux épistémiques et éthiques autour de la pandémie de COVID-19," especially Thomas Bonnin, Samuel Lepine, Juliette Ferry-Danini, Michel Dubois, and Clément Tarantini; and an audience at the Philosophy of Science Association (PSA) 2021 conference, especially Rachel Ankeny, Rachael Brown, and Gregory Morgan. Discussions with Sara Green and Andrew Kilianski on earlier versions of this material were also extremely helpful.

I've also been able to present the broader idea of integrative promise as a whole on a number of occasions; thanks to audiences at Université de Lausanne, especially Christophe Dessimoz, Marc Robinson-Rechavi, Béatrice Desvergne, Alexandre Reymond, Ted Farmer, Tanja Schwander, Yolanda Schaerli, Christian Fankhauser, Anna-Sapfo Malaspinas, and Sakshi Sharda; at ISHPSSB 2021, especially Matt Haber, Greg Priest, Trevor Pearce, Jon Hodge, Rose Novick, Betty Smocovitis, Rose Trappes, and Marshall Abrams; at the 2021 meeting of the Société de Philosophie des Sciences, especially Anatolii Kozlov, Massimiliano Simons, Thomas Reydon, Nicola Bertoldi, and Francesca Merlin; at the UCLouvain CEFISES Work in Progress Seminar, especially Peter Verdée, Cristian Lopez, Kévin Chalas, Maxence Gaillard, Juliette Ferry-Danini, Max Bautista Perpinyà, and Mathieu Berteloot; and at the University of Bielefeld, especially Alkistis Elliott-Graves, Philip Haueis, Martin Carrier, Christian Nimtz, and Rose Trappes.

Informal chats with a huge number of colleagues have also shaped the project; in particular I owe thanks to Markus Eronen, Hugh Desmond, Bendik Hellem Aaby, Dan Nicholson, Erin Saupe, and Adrian Currie. Extra thanks are also due to Juliette Ferry-Danini, who inspired a late-stage change in the very name of integrative promise itself.

Finally, thanks to Otávio Bueno, the series editor for the Synthese Library, as well as Christopher Wilby and Palani Murugesan at Springer, for their support and their patience. The book also benefited from the efforts of an extremely thorough, charitable, and generous reviewer.

Last but assuredly not least, I'll never be able to adequately thank my family. Being an overcommitted academic through a global pandemic with a backlog of projects much too long for my own good has meant that I have relied upon them well more than I should have. Julia (along with Ada, Curie, and the late, and much-missed, Schrödinger) and my parents have kept me sane against all odds.

Brussels, Belgium
February 2025

Charles H. Pence

Contents

Part I
Integrative Promise and the Explanatory Virtues

Chapter 1
Virtues and Pursuit-Worthiness

Abstract This chapter briefly sketches the general concept of integrative promise, describing it as a pattern of invocation of explanatory virtues: in particular, integration, stringency, opportunism, and appeals to non-epistemic values. First, I defend the choice of the somewhat unusual term "explanatory virtue," by comparison with appeals to other, more classic accounts of theoretical virtues, particularly those of Thomas Kuhn. I then situate integrative promise with respect to a number of other existing philosophical literatures which have treated similar subjects. Especially important is the concept of pursuit-worthiness, which has already been the topic of a significant amount of philosophical discussion. I also describe the relationship between integrative promise and the widely recognized theoretical virtues of fruitfulness and scope, and Mayo's concept of severe testing.

Keywords Explanatory virtues · Theoretical virtues · Pursuit-worthiness · Fruitfulness · Scope · Consilience · Severe testing

This book presents an argument for the claim that, across the history and contemporary practice of the life sciences, biologists have favored explanations that exemplify a certain cluster of epistemic and non-epistemic values, a cluster which I will call *integrative promise*. Put differently, when biologists have evaluated explanations—whether they be extant explanations of a given biological phenomenon, or putative future explanations towards which to direct research efforts—they have frequently done so by appealing to the integrative promise of those explanations. Broadly speaking, integrative promise thus refers to an explanation's (putative or demonstrated) ability to unify different classes of biological phenomena. But this is only part of the story; as we will see, such integrations, in part because of their difficulty and in part because of their importance, also involve the invocation of other goals beyond integration itself.

In this introductory chapter, I will pursue two related objectives. First, I will present the general intuition behind integrative promise—light on concrete detail but hopefully suggestive of the motivation for engaging in the project that follows. I will then turn to a discussion of other philosophical concepts related to integrative

C. H. Pence, *Integrative Promise*, Synthese Library 499,
https://doi.org/10.1007/978-3-031-88471-9_1

promise by way of ground-clearing and stage-setting, situating it in the literature on theoretical virtues and pursuit-worthiness. I close by setting up the rest of the first part of the book, identifying the four features that, I will argue, constitute integrative promise.

1.1 The Intuition Behind Integrative Promise

The contemporary life sciences set for themselves a difficult task. Nearly every textbook in an introductory biology course contains something like Fig. 1.1, in which we learn that the biological "levels of organization" range from something like, at one extreme, macromolecules and the other ingredients of biochemistry, to, at the other, the entirety of life on Earth. This is an impossibly large remit—and yet there is a pervasive and justifiable sense in which, in order to understand the living world, we will have to incorporate knowledge of at least a significant number of these domains. As early as 1959, C. H. Waddington wrote that in addition to the

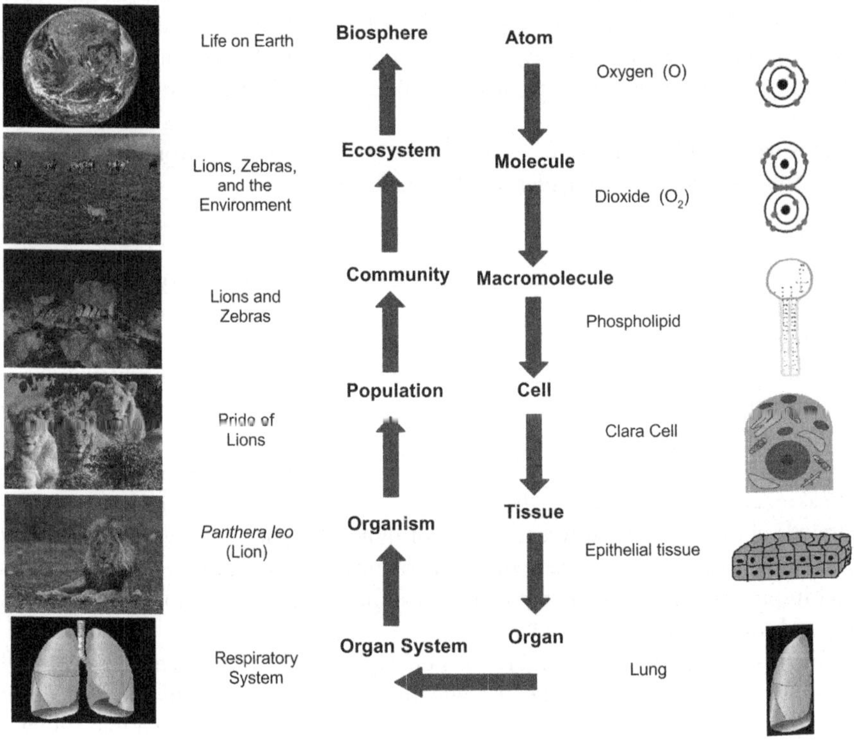

Fig. 1.1 A typical textbook-style diagram of biological "levels of organization," ranging from atoms to the biosphere. Wikimedia Commons, CC-BY-SA 3.0, by Mikala14

integration of genetics with evolutionary theory that had unfolded in the preceding two decades,

> The time seems to have come when we need to take into account two further aspects of the evolutionary mechanism. In the first place, natural selection pressures impinge not on the hereditary factors themselves, but on the organisms as they develop from fertilized eggs to reproductive adults. We need to bring into the picture not only the genetic system by which hereditary information is passed on from one generation to the next, but also the 'epigenetic system' by which the information contained in the fertilized egg is expanded into the functioning structure of the reproducing individual. Each organism during its lifetime will respond in some manner to the environmental stresses to which it is submitted, and in a population there is almost certain to be some genetic variation in the intensity and character of these responses. (Waddington 1959, p. 1635)[1]

Here, more than sixty years ago, we have a call for the coalescence of the cellular/developmental, organismic, populational, and ecosystemic or environmental levels of biological explanation. As we will see unfold across a host of examples in this book, as our capacity to pursue more complex and detailed biological experiments has increased, demands for this cross-domain approach have only intensified.

On the one hand, we could have a somewhat deflationary reading of these discussions. There is assuredly a kind of tacit assumption present in at least some of these cases that there will be a division of labor here—in the same way as has traditionally been the norm for other sciences, individual domains of the life sciences will be isolated and studied independently of one another by groups with largely autonomous research traditions, separate sources of funding, and distinct experimental techniques. We could thus interpret calls like Waddington's as expressing a hope that the *community*, across its various sub-communities, would have the collective ability to understand each of these levels.

But on the other hand, a persistent trend throughout the history of biology has advocated for genuinely *integrating* these levels, not merely accumulating isolated data about each taken separately. Theodosius Dobzhansky's celebrated quip that "nothing makes sense in biology except in the light of evolution" (Dobzhansky 1964, p. 449) elevates one common candidate for the type of perspective that might let us explore all of these levels at once: Darwinian evolution. Historically, others have been offered, from various concepts of organic form (Russell 1917) to understandings of nutrition and metabolism (Bognon-Küss 2019). In all these cases, we have a proposal that the pursuit of certain kinds of explanations holds out hope for a real inter-domain explanation, an integration of knowledge that becomes somehow more than the sum of its parts.

Through the widest lens, the argument that I will develop in this book is that a significant thread throughout past and present work on the life sciences can be illuminatingly traced if we see appeals to the *promise* of such large-scale *integration* between biological domains as an important motivation for biological study. From

[1] For an illuminating discussion of Waddington's use of 'epigenetic' (and changes in the term in the decades since), see Nicoglou and Merlin (2017).

Darwin's day to our own (or, at least, this will be the extent of my analysis here), we find biologists pursuing programs of research because they offer us this *integrative promise*.

Promise in this sense—in what follows, I will often use "promise" as a shorthand for "integrative promise"—will clearly not be a single, clear-cut phenomenon with a precisely definable scope. My task is thus to try to shed light on a *pattern* of evaluation—a recurring use of certain epistemic and non-epistemic values in the pursuit of explanatory goals, or, as I will refer to it for short, a set, pattern, or collection of *explanatory virtues*. This also makes promise philosophically interesting, insofar as it offers us yet another site of important interactions between epistemic values (i.e., biologists pursue promise because they believe it will lead them toward more, and more interesting, truths about the world) and non-epistemic values (i.e., they also pursue promise because it will help them fulfill various kinds of pragmatic and social goals).

The structure of my argument will be firmly rooted in the tradition of integrated history and philosophy of science. Inspired by other recent works in this vein (with a particular hat-tip to Shan 2020), I think that the proof of the pudding is in the eating: the utility of taking integrative promise as a theoretical framework for understanding the history and philosophy of the life sciences is best demonstrated not at the abstract or theoretical level, but in seeing what new insights it can give us as we approach particular concrete cases. From Darwin on orchids to contemporary big data and digital analysis, I contend, taking up promise as a framework for looking at biological theory offers us enlightening perspectives, draws our attention to connections that we might have otherwise missed, and can help us better understand how biologists themselves approach the present and future of their field.

What are the elements that make up this composite explanatory virtue—that is, that comprise this recurring pattern of particular epistemic and non-epistemic values common across the biological sciences? Explicating them will be the project of the rest of the first part of the book. For the moment, therefore, we will have to rest content with a brief sketch. Because promise is a somewhat diffuse pattern, I cannot offer here a traditional definition of the concept, a set of necessary and sufficient conditions for determining whether or not promise has been used in a particular context as a criterion for scientific decision-making. What I will do instead is to tease out four significant dimensions of integrative promise, and illustrate the ways in which particular historical and contemporary examples that manifest a use of promise have invoked these four different dimensions. Cases where we see these values utilized, then, are likely to be those in which judgments of integrative promise are important. Those four axes will form the topics of Chaps. 2–4, and I will then turn in Chap. 5 to a final instance that shows all four in action in a single case.

Integration The first of these is integrative promise's most important epistemic feature: integration. As we will see, the kinds of projects pursued under the banner of integrative promise are often relatively risky. They might involve the development of elusive theoretical resources, or the pursuit of costly and painstaking empirical or observational data collection. This kind of risk must be offset by a meaningful epistemic gain. That gain is integration.

As just noted, biology's status as a science with aspirations to understand levels from the biochemical to the ecosystemic leads to a serious challenge. Any explanatory project that offers up the possibility of integrating these different levels is thus taken to be worth an elevated measure of risk. That said, there is much work to be done to make this notion more precise, which will be the topic of Chap. 2. The very idea of "levels of organization"—and with it, the coherence of diagrams like Fig. 1.1—has been harshly questioned in recent scholarship in the philosophy of biology. I will thus turn to the development of a more general notion of "inter-domain integration" which, I argue, can capture the relatively common intuition found in biological practice that there is *something* being integrated, without committing us to the kinds of problematic metaphysical claims that have been the target of attack.

Stringency and Opportunism I will then turn to just how the particular projects pursued for their integrative promise tend to be selected. First, one apparent factor at work here offers us another distinction between promise and empirical scope (the simple pursuit of a wider field of application for a given theory). When biologists pursue integrative promise, they don't do so by moving toward whatever extensions of scope might be most readily available (and thus they don't demonstrate a search for empirical scope for its own sake). Instead, they advance particular kinds of explanatory projects that will be likely to offer us the kind of (relatively rare) evidence that could nonetheless support a case for integrative promise. I'll call this *stringency.*

But—moving in the opposite direction—we also see that the explanations sought for reasons of promise are tempered with a healthy dose of realism, which I will call *opportunism.* By contrast, for instance, with the pursuit of novel fundamental theory in the physical sciences (what we might think of as the closest analogue in physics of integrative promise), we see rather that these lines of research are often constructed around the points of intersection of our current best theories, or places where there are substantive existing experimental resources, such as protocols and model organisms. This opportunism gives biologists a way to pursue what would otherwise be impossibly difficult integrative research with a reasonable likelihood of success. Invocations of stringency and opportunism are thus important for bringing integrative promise "down to earth," we might say—for ensuring that explanations sought for their integrative promise might actually be realized, and when realized, might actually support the case for integrative promise. I will consider these two criteria in Chap. 3.

Non-epistemic Values Lastly, and perhaps most importantly, I will argue in Chap. 4 that judgments of integrative promise are a particularly important site of interaction between epistemic values, such as integration, borderline values that are neither clearly epistemic nor non-epistemic (such as opportunism), and non-epistemic or social values. Very often, appeals to promise are packaged with calls for biology to meet large, socially desirable "grand challenges:" think of the hype surrounding personalized medicine at work in discussions of the Human Genome Project, or recent emphasis on biology's role in the response to COVID-19.

Of course, one could argue that these are merely cynical appeals: after all, integrative explanations are likely to be the fruit of expensive research projects, and raising funding for such pursuits becomes easier when one can tie them to social goals that are likely to attract political support. But I think this cynicism would be too quick: the biologists involved in these appeals really do genuinely believe that part of the integrative promise of these research projects is their ability to advance social and cultural ends. For those of us interested, then, in how non-epistemic values find themselves expressed in contemporary scientific practice (in the broader tradition of recent science-and-values work such as Brown 2013; Soler et al. 2014; Elliott and Richards 2017), appeals to integrative promise can give us a particularly fruitful place to analyze this relationship.

A full introduction of each of these facets of promise and their interaction will have to wait for the chapters to come. For now, however, I should pause to explore a few connections with similar ideas already familiar from the literature in the philosophy of science.

1.2 Promise and Neighboring Notions

Readers confronted with the basic presentation of integrative promise above will likely have already noted that this is well-traveled philosophical terrain, in a variety of senses. Building on Thomas Kuhn's initial introduction of the criteria of theory change as *values* (Kuhn 1977, 1996) and the separation of those values into categories of epistemic and non-epistemic by Ernan McMullin (1982), there is a substantive and flourishing literature on what are now called *theoretical values* or *theoretical virtues*. While I will turn to a more detailed discussion of theoretical virtues in a few moments, their importance can be underlined by returning to Kuhn's classic presentation of their role in the choice between theories. Faced with a decision about which theory to adopt, a scientist evaluates the available candidates according to a number of different criteria. But far from being straightforward, timeless characteristics of rationality, or the like, these criteria, as Kuhn argued, function rather as values—in the sense that scientists might not only differ among themselves about the relative importance of the criteria found on the list, but they also might differ about the extent to which any one theory exemplifies (or fails to exemplify) any one such criterion.

In addition to this literature on theoretical virtues, any appeal to *promise* calls to mind discussions of what philosophers have termed *pursuit-worthiness*. Another current throughout the literature on theory choice, introduced in its contemporary sense by Larry Laudan (1977), considers how the choices of scientists are not only shaped by the virtues currently exhibited by theories (that is, the extent to which a theory might be supported by today's data, or presently free of contradiction), but also by how scientists believe that those theories will continue to change as they develop—by the extent to which they are worthy of further pursuit.

It is important to underline, therefore, that integrative promise is not an entirely novel notion. Rather, I want to argue that it is intimately related to several extant understandings of theoretical virtues and pursuit-worthiness. I will take integrative promise to be a *pattern* of appeals to certain epistemic and non-epistemic values, as applied to scientific explanations, and those values are closely allied to those already explored elsewhere. What makes it an interesting phenomenon is the fact that this pattern emerges, repeatedly, across a variety of domains and a variety of historical and contemporary cases in the life sciences. In that sense, I will argue, exploring promise in itself is a useful enterprise insofar as it helpfully illuminates something that these widely diverse examples have in common.

In the rest of this chapter, I'll explore those related philosophical concepts in the following way. First, astute readers will notice that I have referred to promise as built from explanatory virtues, a term that is not particularly common in the philosophical literature. I'll distinguish, in Sect. 1.3, theoretical from explanatory virtues, and argue for why I believe promise is better understood as the latter. In Sect. 1.4, I discuss the relationship between promise and established theoretical virtues—particularly, fruitfulness, scope, and (to take a somewhat unusual theoretical virtue) consilience. I'll compare the idea of stringency with Deborah Mayo's concept of severe testing in Sect. 1.5. I'll then consider, in Sect. 1.6, the relationship between promise and pursuit-worthiness. Pursuit-worthiness, I will argue, is more general than integrative promise, which, by being restricted to certain particular cases, can reveal something special about the way in which the life sciences operate.

1.3 Theoretical and Explanatory Virtues

As already mentioned above, we can profitably begin a discussion of theoretical virtues (or theoretical values) with Thomas Kuhn.[2] In 1969, seven years after the initial appearance of *The Structure of Scientific Revolutions*, Kuhn added a postscript, intended in large part to blunt criticism that he had received concerning his supposed radical, relativist tendencies—tendencies which he would disavow for the remainder of his career.

Kuhn is at pains to note that the kind of relativistic anarchy that one might read into the pages of *Structure* is merely apparent. Decisions between theories are governed by values. These values, while they do not leave us with a singular, rational

[2] I do not believe that anything significant hinges upon the distinction between values and virtues in this context. In both cases, I intend to refer to a criterion of judgment or evaluation, which may be exemplified to varying degrees, and about the application and importance of which rational agents of good faith may disagree. Any further distinctions between these two notions are unimportant for my project here.

choice procedure for deciding between theories, nonetheless are a far cry from the kind of "anything goes" approach that would soon be wholeheartedly embraced by Paul Feyerabend (1975).

As he presents the argument a few years later, in a lecture published as part of *The Essential Tension*, we begin with the very notion of criteria for theory choice:

> What, I ask to begin with, are the characteristics of a good scientific theory? Among a number of quite usual answers I select five, not because they are exhaustive, but because they are individually important and collectively sufficiently varied to indicate what is at stake. (Kuhn 1977, p. 321)

I will return to Kuhn's list of criteria in the next section—for now, let's stay with the nature of these characteristics themselves. What kind of thing are they, and how do they work, if, following Kuhn's earlier argument, they are not to give us a cookbook for an objective choice between competing theories?

In short, Kuhn writes that "the criteria of choice with which I began function not as rules, which determine choice, but as values, which influence it" (Kuhn 1977, p. 331). A discussion of values first enters in the postscript to *Structure*, as Kuhn tries to clarify just what it is that makes up a paradigm, or, in his later vocabulary, a disciplinary matrix. Values, he writes, "are more widely shared among different communities" than the traditionally evaluated aspects of scientific theories, "and they do much to provide a sense of community to natural scientists as a whole" (Kuhn 1996, p. 184). What makes us think that certain parts of paradigms function as values? In Kuhn's words,

> To a greater extent than other sorts of components of the disciplinary matrix, values may be shared by men who differ in their application. ... judgments of simplicity, consistency, plausibility, and so on often vary greatly from individual to individual. [...] Even more important, in those situations where values must be applied, different values, taken alone, would often dictate different choices. (Kuhn 1996, p. 185)

But this need not mean a kind of relativist anarchy in the scientific process. Two people committed to the same set of values might make different choices in a particular setting. Values themselves may be ambiguous, and may be prioritized in a variety of different ways. But they can nonetheless function as, in some sense, constitutive of what it is to be doing science, and we can have profitable, rational arguments with one another in an attempt to settle these kinds of value disagreements (Kuhn 1977, p. 331).[3]

At least this much has become a fairly broad consensus in the philosophy of science: these criteria for theory choice do not function as hard and fast rules, but rather more like values, which constitute in the end something resembling, as Kuhn also sometimes put it, a flexible heuristic rather than a hard-and-fast algorithm for selecting which theory to adopt. For my purposes, this is complexity enough—we can rest content with the relatively thin definition of theoretical virtues that Samuel Schindler offers us, when he asks: "What are the characteristics of a theory which

[3] As he writes, adding or removing values will be likely to change the very nature of the enterprise in which we are engaging—adding social utility might turn science into engineering. As we will see in Chap. 4, I think this is too quick, if nonetheless a step in the right direction.

scientists value and which guide them in their choice to adopt one theory or another? In other words, what are the virtues of a scientific theory?" (Schindler 2018, p. 5).

While these are taken to be virtues of *theories* as a whole, it is fairly easy to see how we might extend such a notion to virtues that apply to *explanations*. Thus, I will define *explanatory virtues* as those criteria which explanations (or potential explanations) exemplify, that lead to such explanations being preferred (or pursued) by scientists.

First, let me explore a few characteristics of this definition, before moving on to a brief argument for why we would be interested in developing such a notion in the first place. Note that this is a definition taken to apply not only in a "backward-looking" sense, to explanations that have already been offered of phenomena, but in a "forward-looking" sense as well, to the pursuit of as-of-yet undiscovered explanations in a given domain. As Eli Lichtenstein has compellingly argued, many disagreements in science that *appear* to be conflicts over the present characteristics of theories are "in fact more directly rooted in divergent assessments of pursuit-worthiness, concerning which theory will be best down the line" (Lichtenstein 2021, p. 174). This seems to be all the more true when we consider evaluations of the quality of explanations. Often, passing judgment on a particular explanation for a given phenomenon is not just about the nature of the phenomenon as we have observed it, or the data gathered so far, but also on the approach used to construct the explanation in the first place—the methodology, the way in which the research question has been framed, or the allocation of resources to the project. These kinds of evaluations seem not only to support or condemn an explanation of one particular result, but also to advance or undermine the case for whether or not any similar such explanations should be expected to succeed in the future. Explanatory virtues, therefore, will need to take these forward-looking dimensions seriously.

Second, theoretical virtues and explanatory virtues will of course be tightly related. Theories that are prized because they offer us accurate predictions, for instance, will almost certainly also offer us accurate explanations of phenomena as well (a claim related to traditional studies of the explanation-prediction connection; see Douglas 2009). Simpler theories might be simpler precisely because they offer simpler explanations. Perhaps most generally of all, it seems that all else being equal, we will tend to prefer theories that produce explanations that bear explanatory virtues, and thus a connection between these two kinds of virtues can be expected *a priori*.

This connection will enable us to understand how integrative promise might be related to examples of theoretical virtues which have already received extensive discussion in the philosophical literature. As I will argue, integrative promise is not precisely captured by any of these common examples, and this constitutes one of the major reasons that I think it deserves its own, independent account.

Now, let's take a step back. Why is it that integrative promise should be expressed as composed of explanatory virtues in the first place, rather than theoretical virtues? A core element of this argument will have to wait for Chap. 6, where I contend that explanations deployed in contexts of contemporary "big data biology" often seem to be evaluated for their integrative promise, even in cases where the very scientists making those evaluations claim that their explanations "transcend" theory or take place in a context "without" theoretical commitments. Of course, one may (and, I think, should) doubt those claims, but it's not clear that it is reasonable to reject them out of hand by making promise only applicable to theories. Put differently, I don't believe that our account of the virtues that biologists are hoping to defend in these contexts should settle that argument *a priori*.

More importantly, explanatory and theoretical virtues are targeted at different contexts. As has already become clear, discussion of theoretical virtues (and, as we will see, of pursuit-worthiness as well) largely arises from the problem of theory choice. We imagine a situation in which a scientific community is choosing which theories to adopt, and hence theoretical virtues enter as criteria for the resolution of scientific disagreement. Judgments of promise, however, often occur in the total absence of inter-theoretic disagreement of this kind. Scientists working entirely within a particular paradigm evaluate the promise of current research projects, as well as propel their future efforts by appealing to their promise—and this without any intention of replacing their current theoretical framework.[4]

One might respond that even if theoretical virtues tend to be evaluated and compared only at moments of theory choice, those virtues are still possessed by theories at other times.[5] To be sure, this is true. But I think that the difference in the functional role played by those virtues—a difference implicitly accepted even by the objection raised here—is important. The fact that explanatory virtues are evaluated within the course of "normal science," to borrow the Kuhnian turn of phrase, while theoretical virtues usually are not, marks out a good reason to think that these virtues are properties of explanations rather than theories.

That said, theoretical disagreement and disagreement over explanatory quality could blur into one another in some cases. To take just one example, choosing to pursue or to avoid a certain explanation for reasons of its promise might amount, in effect, to either creating or abandoning a particular theory—a phenomenon of which we'll see several examples in the chapters that follow. Explanation and theory construction, unsurprisingly, cannot be entirely divorced in practice. But the existence of these edge cases does not imply, I think, that promise is constituted of theoretical rather than explanatory virtues.

To sum up, then, I have argued for the existence of a new category of analysis, *explanatory virtues*, by analogy and in close dialogue with theoretical virtues. A motivated skeptic could still resist this move, arguing that explanatory virtues are simply a way to view the theoretical virtues of the theories that have produced those explanations—that theoretical virtues, in some sense, "trickle down" to the explanations that those theories create. I have offered a few justifications for why I think ascribing these virtues to explanations is more perspicuous and clearer, but little will turn on this point in what follows, so I hope the skeptic will continue to follow along with the project.

---

[4] Precisely the same critique will be applicable to the literature on pursuit-worthiness, to which I will turn in Sect. 1.6.

[5] Thanks to an anonymous reviewer for posing this objection.

As I'll describe it in the remainder of the book, then, integrative promise is best understood as a pattern of explanatory virtues. I'll present a collection of virtues of biological explanations that tend to "travel together" and to be defended as a unit in biological practice. I don't mean to propose that promise exhausts the entire category of explanatory virtues—there certainly could be others! But the four aspects that I contend constitute promise are my analytic target here; investigating further explanatory virtues, or the potential relationship between other theoretical virtues (beyond those I will discuss in the next section) and explanatory virtues is a task for future work.

1.4 Promise and the Other Virtues

As just noted, it is clearly the case that theoretical and explanatory virtues will be related to one another, and thus that we can begin to situate a discussion of integrative promise in the context of the literature on theoretical virtues. As we will see, my argument is that promise deserves independent treatment not because it is entirely novel, but because it captures a repeated appeal to a set of explanatory virtues that are, in turn, related to a number of different theoretical virtues already considered in the literature—something like a pattern of invocations of those other epistemic and non-epistemic values where the whole is more than the sum of its parts. I'll here consider three theoretical virtues with important relationships to promise: fruitfulness, scope, and consilience. The first of these is relevant to what we might call promise's "future-directedness," while the latter two are related to integration.

1.4.1 Fruitfulness

In his initial list of theoretical virtues in *The Essential Tension*, Kuhn mentions accuracy (i.e., fit with existing data), consistency (both internal or logical and with other theories), scope (about which more in the next section), simplicity, and fruitfulness (Kuhn 1977, pp. 321–322). He places particular emphasis on this last virtue, sometimes also called fertility or fecundity:

> Fifth—a somewhat less standard item, but one of special importance to actual scientific decisions—a theory should be fruitful of new research findings: it should, that is, disclose new phenomena or previously unnoted relationships among those already known. (Kuhn 1977, p. 322)

He continues, in a footnote:

> The last criterion, fruitfulness, deserves more emphasis than it has yet received. A scientist choosing between two theories ordinarily knows that his decision will have a bearing on his subsequent research career. Of course he is especially attracted by a theory that promises the concrete successes for which scientists are ordinarily rewarded. (Kuhn 1977, p. 322, n.)

The literature on fruitfulness is rather complex and, as we will see also in the case of scope in the next subsection, has a tendency to muddle a variety of different senses of the term. Kuhn here seems to invoke both a sort of epistemic fruitfulness (that the theory indicates the presence of previously unknown phenomena or connections) as well as a pragmatic fruitfulness (that adopting the theory will prove good for one's career, publications, grants, and so forth). Unfortunately, Kuhn doesn't make this notion any more systematic elsewhere.

Prior to the development of related discussions of pursuit-worthiness, which I'll take up separately in Sect. 1.6, perhaps the most detailed account comes from Ernan McMullin (1976, 1982). In his well-known work drawing the distinction between epistemic and non-epistemic values, he briefly considers two potential ways to flesh out fruitfulness. First, a theory could be fruitful in the sense that it "proves able to make novel predictions that were not part of the set of original explananda" (McMullin 1982, p. 16). This meaning will not be important for us here, despite the fact that it has to some extent become the standard notion of fruitfulness in the literature, especially as it has been taken up in the realism/anti-realism debate. Schindler, for instance, notes that "nowadays, theoretical fertility is normally understood in terms of novel success" (Schindler 2017, p. 151). When I talk about integrative promise, however, I do not mean finding results that were not expected at the time the theory was proposed—I rather have in mind the targeting of particular kinds of goals, desirable for other reasons, which scientists have reason to think can be attained by research in a particular direction. The type of "fruitfulness" invoked by integrative promise is not an epistemic credential, but a gamble on the future, and in particular a gamble on the potential to obtain a specific kind of (integrative) result. As we will see in the case of pursuit-worthiness, we are at best after a subset of fruitfulness in McMullin's first sense.

McMullin's second notion of fruitfulness does not do much better for our purposes. We might prefer a theory if it "proves to have the imaginative resources, functioning here rather as a metaphor might in literature, to enable anomalies to be overcome and new and powerful extensions to be made" (McMullin 1982, p. 16). As Schindler elaborates the notion, this is a diachronic property, a demonstrated track record of a theory to make the sorts of modifications required to respond to unexpected experimental or observational results (Schindler 2017, pp. 153–154). Again, however, we have a theoretical virtue without much relationship to integrative promise. As we will see in a number of examples throughout the rest of this volume, many of the theories that are pursued for their promise do not have this kind of track record to appeal to—indeed, they may not have any track record at all.

While there is certainly a link between integrative promise and fruitfulness, it does not seem to be fruitfulness in any of the senses that have been elaborated so far in the literature. More broadly, this makes sense. Analyses of fruitfulness as a theoretical virtue have attempted to interpret it as an epistemic value—that is, as something that is valued because theories which bear it (i.e., which are fruitful) are more likely to be true. While, as we will see especially in Chap. 2 on inter-domain integration, there is a clear and important epistemic aspect to the ascription of promise, the promissory note for future success developed in the context of integrative promise is not taken to be an epistemic criterion. Many of the pursuits launched under the banner of integrative promise might even be viewed as particularly dangerous from the viewpoint of fruitfulness of the traditional sort—risky gambles that are sought for reasons that go beyond their demonstrated epistemic credentials. As we will see emerge over the next several chapters, the case for this future success thus has to be advanced by a complex interaction between various epistemic and social value judgments.

1.4.2 Scope

The fact that my discussion of integrative promise places such importance on the notion of inter-domain integration makes for an obvious comparison with another of the classic Kuhnian theoretical virtues: scope. Scope makes for an interesting dilemma—despite its appearance in nearly every list of theoretical virtues, it is almost never discussed explicitly, and thus there is very little philosophical guidance for what scope might actually mean.[6] Kuhn himself writes that scope amounts to the claim that "a theory's consequences should extend far beyond the particular observations, laws, or subtheories it was initially designed to explain" (Kuhn 1977, p. 322). Notably, this is almost coextensive with the definition of fruitfulness as novel prediction that we saw (and dismissed) above in the guise of McMullin's first definition of fruitfulness.

Perhaps the most common concept of scope phrases it instead in terms of *unifying power*. In fact, on McMullin's list of epistemic values, unifying power replaces scope, as "the ability to bring together hitherto disparate areas of inquiry" (with the examples of Maxwell's electromagnetism and, more unusually, plate tectonics; McMullin 1982, p. 15). Schindler makes the same move, arguing that "a theory has unifying power (or broad scope) when it unifies phenomena which were previously considered distinct" (Schindler 2018, p. 11).

What exactly do we mean when we talk about unifying power or theoretical unification in this sense? A revealing illustration can be drawn from the work of Philip Kitcher on unification as the basis of scientific explanation.[7] He constructs his concept of unification around the illuminating example of Newtonian mechanics. In the days after Newton had shown that celestial and terrestrial movement could be unified under a relatively compact set of supposedly universal laws, the Newtonian program held out the possibility of bringing together vast domains of science under a similar set of explanations phrased in terms of inertial conditions and perturbing forces (Maudlin 2004). Calling the approach "dynamic corpuscularianism," Kitcher

[6] A notable exception is Angela Potochnik's (2017, pp. 140–141) discussion of scope in the context of causal patterns, an interesting approach that is nonetheless too far afield for my project here.

[7] Potochnik traces philosophical work in this vein even farther back, to the work of Neurath (Potochnik 2017, pp. 188–189).

states that "the next step would be to isolate a few basic force laws, akin to the law of universal gravitation, so that, applying the basic laws to specifications of the dispositions of the ultimate parts of bodies, all of the phenomena of nature could be derived" (Kitcher 1981, p. 513). As Newton put the project in the preface to the second edition of the *Principia,*

> If only we could derive the other phenomena of nature from mechanical principles by the same kind of reasoning! For many things lead me to have a suspicion that all phenomena may depend on certain forces by which the particles of bodies, by causes not yet known, either are impelled toward one another and cohere in regular forms, or are repelled from one another and recede. Since these forces are unknown, philosophers have hitherto made trial of nature in vain. (Newton 1999, pp. 382–383)

Not only the phenomena of physics, but all the phenomena of the sciences thus stood as potential candidates for this unifying effort—and, by extension, the expansion of the scope of the Newtonian paradigm.

This didn't serve as an idle exhortation. As Theodore Brown has noted, a whole group of "Newton-struck" physicians attempted to build force-based approaches to medicine in the late seventeenth century. Archibald Pitcairne, a Scottish doctor, developed "a scheme of 'Postulates,' 'Definitions,' and 'Axioms' from which he claimed to derive a rigorously complete system of unassailably true medicine" (Brown 1987, p. 632). "A Living Body," he wrote, connecting together several of his definitions, "is compounded of Canals of diverse kinds, conveying different sorts of Fluids" (Brown 1987, p. 632). Theoretical unification under the Newtonian banner was thus seen not only as an ideal, or merely as philosophically desirable, but as a live possibility, to be pursued as an active aim of scientific research.

There is, I think, an important sense in which theoretical unification is closely related to one of the explanatory virtues that makes up integrative promise. As we will see in Chap. 2, integration and unification have long been linked—in opposition to one another!—in the literature on integration in the philosophy of science. And in a colloquial sense, a "desire to unify" will certainly be found in some of the examples of integrative promise that I will discuss in what follows. When we turn, for instance, to the discussion of Darwin's late work in Chap. 3, we will see that part of Darwin's motivation is assuredly the desire to unify a wide array of phenomena under a kind of overarching evolutionary explanation. But two things lead me to think that integrative promise is not *just* a matter of theoretical unification. First, as already discussed above, insofar as theoretical unification is a theoretical virtue, rather than an explanatory one, my arguments in the last section that integrative promise is composed of explanatory virtues can be duplicated here.[8]

Second, as I have attempted already to indicate and will discuss in much more detail in what follows, integrative promise is not merely an epistemic value.

[8] It is an interesting possibility (that I lack the space to explore here) that theoretical unification might itself also be best construed as an explanatory virtue—it consists in the search for explanations of particular phenomena that bring those phenomena within the scope of another theory. This is not, however, the way in which it is commonly presented in the literature; a theory is rather said to exhibit unifying power if it can generate such explanations.

Unifying power is certainly one part of the epistemic portion of integrative promise, but will struggle to encompass the pragmatic- and social-value sides to promise that I will describe in what follows. Unifying power is thus an important point of comparison and context, but only a piece of the larger picture of integrative promise.

Before leaving scope entirely, we should pause to make one further distinction, found in Schindler's work on theoretical virtues, which will be important for us later. As he notes, in addition to scope in the sense of unifying power, there is another way in which a theory might be said to have a broad scope:

> ...we should note that unifying power should be distinguished from the *empirical* scope of the theory. A theory can conjoin many facts and therefore have broad empirical scope but little unifying power. That would be the case if the theory gave us no clue as to how the conjoined facts are interrelated at a 'deeper' level. (Schindler 2018, p. 12)

Dividing the more general concept of scope into *empirical scope* and *unifying power* offers us a helpful conceptual tool, and one that I will use occasionally throughout the book.

1.4.3 Consilience

While it is not a standard feature on lists of theoretical virtues (it is, for instance, absent from Schindler 2018), integrative promise is also reminiscent of *consilience*, another traditionally appreciated feature of scientific theories. Coined by William Whewell (though referring to a concept present throughout the history of the philosophy of science), the "consilience of inductions" is one of the central elements of his approach not only to the philosophy of science, but also to his presentation of the history of science, which, he argues, shows us a progressive chronicle of expanding consiliences.

> *The Consilience of Inductions* takes place when an Induction, obtained from one class of facts, coincides with an Induction, obtained from another different class. This Consilience is a test of the truth of the Theory in which it occurs. (Whewell 1847, p. 469)

While unpacking the definition of the concept as it appears in Whewell's own work is not a simple matter, and would take us too far afield for my project here (see Laudan 1971; Quinn 2017), the idea has taken on a kind of standard meaning in the years since its inception: we have a particularly strong argument in favor of a given theory if multiple, independent lines of reasoning all offer support for it.

To be sure, one might simply interpret this as a kind of argument in favor of a theory's acceptance after it has been proposed; Whewell's description of consilience as a "test of the truth" of a theory seems to push us in this direction.[9] That said, there is also a way of reading consilience that appears to be more general than this, and hints of it can even be found in Whewell's own writings. If an explanation

[9] I thank an anonymous reviewer for raising this objection.

originally proposed for one class of phenomena, "of itself and without adjustment for the purpose, gives us the rule and reason of a class of facts not contemplated in its construction" (Whewell 1847, pp. 67–68), then I think we're entirely justified in seeing this as a virtue of that explanation—it has enabled us to arrive at a consilience, and its plausibility is therefore supported by multiple lines of evidence, making it a better explanation of the phenomena at issue than one that lacks this property.

In this sense, then, consilience bears an interesting relationship to promise. If an explanation bearing integrative promise can be carried through, then it will, practically by necessity, be supported by a consilience—that is, if it successfully offers us a way to provide inter-domain integration, then it will be supported by multiple different types of evidence.[10] But because of the future-directed nature of promise, this kind of consilience cannot be demonstrated in advance. Consilience could thus be seen as an aspirational goal for the pursuit of integrative promise; if we succeed in developing the kind of explanations for which promise tells us to search, they will in turn function as consiliences supported by a wide variety of biological evidence.

1.5 Severe Testing

Let me now turn to a close relative of the stringency criterion that is familiar from elsewhere in the literature in philosophy of science.[11] Deborah Mayo, in her extensive contributions to debates in the philosophy of statistical inference, has proposed a notion that she refers to as *severe testing*. Given that I've introduced stringency above as the search for the rare sorts of evidence that could support arguments for integrative promise, this might naturally be read as an appeal to something like a theory's ability to survive severe tests. Should we see stringency as severe testing in another guise, or as something different?

To explore the relationship between these two concepts in greater detail, it is important to start by understanding just what severe testing is and in what context Mayo deploys it. Mayo's intent is to provide us with an understanding of statistical tests on which certain uses of those tests—carefully governed by a series of norms for statistical practice—can offer genuine, positive reasons to believe in the truth of a theory. This means, among other challenges, that she needs to respond to many of the classic objections that were raised against other theories of confirmation over the course of the twentieth century. Among these is a kind of generalized Duhem-Quine problem of the underdetermination of theories by the available evidence.

[10] One finds similar arguments in contemporary discussions of scientific pluralism (Kellert et al. 2006).

[11] My thanks to an anonymous reviewer for mentioning this parallel, which had escaped me while drafting earlier versions of the book.

Put briefly, if, for any theory that might be confirmed by a given statistical test, there are a whole host of other, competing theories that could be equally well confirmed by that same test, then we lack any reason to prefer our current theory over the available alternatives. Why think that there are such competing theories? Well, imagine taking the theory that we're currently testing and reformulating its mathematical foundations in a new way. Or add a few ad-hoc but innocuous conceptual assumptions to the theory. In these and any number of other ways, we could imagine producing a whole host of alternative hypotheses. As Mayo briefly sums it up, "evidence in accordance with hypothesis H cannot really count in favor of H, it is objected, if it counts equally well for any number of (perhaps infinitely many) other hypotheses that would also accord with H" (Mayo 1996, p. 174).

Her requirement for severe testing enters precisely as a response to the threat of this problem of underdetermination. For even if we agree with the worry that there might be a number of equivalent hypotheses, any of which could be supported by a given statistical test, it is unlikely that, for all of those hypotheses, that same statistical test would serve as a *severe test* of the hypothesis—a test that the hypothesis would have been very likely to fail had it been incorrect. As she puts it, when comparing H with an alternative H',

> Although H' may accord with or fit the evidence as well as H does, the fact that each hypothesis can err in different ways and to different degrees shows up in a difference in the severity of the test that each can be said to have passed. The same evidence effectively rules out H's errors—that is, rules out the circumstances under which it would be an error to affirm H—to a different extent than it rules out the errors to affirming H'. (Mayo 1996, p. 212)

I take it that this notion is related to but not identical with stringency, in part because Mayo's approach is entirely cashed out within the context of her views on statistical testing or error statistics (she sometimes calls the notion "error severity" instead; Mayo 1996, p. 178). Let me draw some (hopefully illustrative) comparisons.

First, of course, the aspect of comparison between multiple hypotheses (or, in our case, explanations) is absent in the case of stringency. In the evaluation of stringency, we aren't concerned about how well a given explanation does *relative to* any other potential explanations of the same phenomenon. What we care about, instead, is that we have demonstrated that we have gone out of our way to find a *domain of testing* that is particularly difficult. This does make for a severe test in one particular sense—insofar as integration would be likely impossible were our explanation not a viable one. Note, though, that it is not falsity, inadequacy, or rejection of our theory that is the potential failure mode, but rather the inability to successfully integrate a domain of inquiry that could, if the explanation had been *more* promising, perhaps have been successfully integrated. Again, this relates to a difference in the target of my work and of Mayo's: while she is interested in constructing a robust theory of theory confirmation by statistical testing, I am not particularly concerned here about whether or not a stringent application of an explanation actually gives us reason to think that this explanation is true. It gives us, instead, reason to think that such an explanation exhibits integrative promise.

Secondly, as we will see in more detail in Chap. 3, stringency is not only about an evaluation of the empirical status of an explanation, or the relationship between the explanation and scientific phenomena. Stringency can be in part set by the social context in which an explanation is being proposed and defended. To anticipate the example that I will discuss in detail later on, Darwin's theory of evolution had always taken botany as a particularly important domain of application, and some of Darwin's early defenders, as well as early skeptics, had been botanists. Extensions of the theory of evolution to the plant kingdom, then, as Darwin did in his works on orchids and insectivorous plants, is a *socially* stringent test. He is directly attempting to convince a powerful audience, and the costs of failure would have been significant in terms of the biological community's view of natural selection's integrative promise.

In short, I think that Mayo has pointed out something essential—that severity of testing is a genuine scientific virtue—and I think that the concept of stringency that I lay out here is a more general notion than her error severity, albeit one that begins from many of the same intuitions.

1.6 Pursuit-Worthiness

Last, but definitely not least, we should turn to the relationship between integrative promise and pursuit-worthiness. As I've presented it thus far, the future-directed nature of promise seems to invoke precisely evaluations of whether or not particular explanatory programs are worth undertaking. In this section, I will offer a brief overview of the literature on pursuit-worthiness, and argue that, indeed, judgments of promise are a certain specialized kind of judgment of pursuit-worthiness. That is, when a scientist pursues a theory for reasons of promise, they are deeming it pursuit-worthy *and* declaring that it has a number of other kinds of explanatory virtues as well.

The contemporary literature on pursuit-worthiness commonly traces the concept back to Laudan, who noted that, despite the importance usually accorded by traditional philosophy of science to the *acceptance* or *rejection* of scientific theories, there is also another context in which theories are evaluated: that of *pursuit*. Beginning from an insight for which he credits Paul Feyerabend, Laudan observes that scientists very often have made the choice to pursue theories "which were patently less acceptable, less worthy of belief, than their rivals" (Laudan 1977, p. 110). There are thus four cognitive attitudes that a scientist might take toward a theory: "acceptance, rejection, pursuit, and non-pursuit" (Laudan 1977, p. 119). Laudan offers us one very specific understanding of what pursuit-worthiness might consist in, defined, in a very Kuhnian vein, as the rate of progress in problem-solving that a theory or research program exhibits. He argues that many of the cases in which scientists work with theories that seem less well supported than their alternatives amount to situations where a loss in overall support is compensated for by improved progress, that is, a more rapid rate of advancement—making such theories a good rational gamble, at least until and unless that rate of change starts to decline.

In the early 1990s, a number of philosophers returned to this suggestion and have since cultivated a thriving literature, expanding upon the somewhat limited connection that Laudan drew between pursuit-worthiness and progress. Peter Achinstein, linking pursuit-worthiness to Charles Sanders Peirce's effort to develop a *logic* of scientific pursuit, notes that a variety of criteria might be appealed to in order to justify pursuit: he mentions general methodological criteria (like simplicity or testability) as well as practical or pragmatic factors (like costs in terms of time, money, or energy; Achinstein 1993, pp. 92–93). Laurie Anne Whitt, connecting pursuit-worthiness to McMullin's discussion of fruitfulness, already considered above, discusses both "historical" and "formal" indicators of pursuit-worthiness, including among the former assessments of how a theory has responded to prior challenges to its accuracy or consilience and among the latter the quality of the analogies and experimental strategies that the theory offers us (Whitt 1992).

Recent work, building on the tradition of analyzing the relationship between science and values, has attempted to offer generalizations of pursuit-worthiness that will be particularly amenable to my presentation here. Šešelja et al. (2012), presenting a schematic definition that will be my primary approach to pursuit-worthiness, express it as the following biconditional:

> It is rational for Y to pursue X if and only if pursuing X is (sufficiently/most/etc.) conducive of the set of goals Z. (Šešelja et al. 2012, p. 53)

By varying the values of X, Y, and Z, we can thus examine the assorted notions of pursuit-worthiness on offer.[12] The authors immediately turn to a number of general clarifications of the idea.

First—and again, paralleling my discussion here—the goals Z are now considered not only to be epistemic (as they were in Laudan's discussion of the rate of problem-solving progress), but also to explicitly include non-epistemic values. "More recent discussions of pursuit," write Šešelja et al. (2012, p. 54), "have taken the perspective of practically relevant goals, where the evaluation in the context of pursuit consists of both epistemic and non-epistemic criteria." In their closely related introduction of "inquiry tickets," DiMarco and Khalifa (2019, p. 1017) write that consideration of pursuit-worthiness lets us see a way in which "the role of non-epistemic values need not be restricted to belief or acceptance in order to be

[12] As noted, I will largely restrict my discussion here to the theory laid out by Šešelja et al. (2012), as its flexibility in terms of the target of pursuit makes it easy to transform into an explanatory virtue. Pursuit-worthiness has, in recent years, been the object of an increasingly large literature. Interested readers can, for instance, consult a recent special issue of *Studies in History and Philosophy of Science* to get an idea of this breadth, including: skeptical takes on the very concept (Shaw 2022), on critique as the inverse of pursuit-worthiness (DiMarco and Khalifa 2022), political aspects of pursuit-worthiness (Baker 2022), or the relationship between creativity and pursuit (Sánchez-Dorado 2023), among others. I take it that, for present purposes, the generality of the approach that I'll follow here will make the claims I draw about the relationship between promise and pursuit applicable to a wide variety of different theories of pursuit-worthiness.

relevant to hypothesis choice." As I will argue in more detail in Chaps. 3 and 4, this is a particularly important feature of integrative promise, which heavily modulates pursuit by the influence of a few important non-epistemic values.

Second, the framework of Šešelja et al. is readily adaptable to explanations (and hence to use as an explanatory virtue) by varying the value of the target *X*. They note that, at the very least, we can consider both "the pursuit of scientific theories and other types of pursuit, such as those regarding epistemic objects or technological developments" (Šešelja et al. 2012, pp. 60–61). Slightly expanding their focus, then, I will take the target *X* to be scientific explanations.[13]

I should also mention another feature of their view, if only to dispense with it. In the examples that I will develop throughout the rest of this work, I will usually cite arguments made by practicing scientists to justify their pursuit of particular kinds of explanations in terms of integrative promise. I will thus almost always take the value of the variable *Y* to be an individual investigator. As Šešelja et al. note, however, this doesn't have to be the case. While pursuit-worthiness at the individual level might be considered to be a kind of "research directive," "communal pursuit worthiness can be given in the form of an evaluative stance," a way to express the corporate judgment of the scientific community concerning whether the pursuit of a given target *X* would be "in the epistemic interest of science" (Šešelja et al. 2012, pp. 71–72). While I think a social account of integrative promise that explored more thoroughly the role that it plays in the biological community at large would be interesting and important, that is not my project here.

Notably, the presentation I've offered—as well as the literature on pursuit-worthiness and theoretical virtues as a whole—has a tendency to draw a bright line between judgments of pursuit-worthiness and other judgments of theoretical virtue. It's not clear, however, that this is a distinction that makes a genuine difference. As already discussed, Whitt (1992) made direct connections between her notion of pursuit-worthiness and McMullin's work on fruitfulness. DiMarco and Khalifa (2019) claim that it is values which themselves directly function as inquiry tickets, as does Lichtenstein, who argues that "it is fair to conclude . . . that criteria for problem salience and methodological approach are most realistically viewed as values or heuristics, rather than as precise rules dictating singular choices" (Lichtenstein 2021, p. 174). It is thus plausible to conclude that the values that constitute pursuit-worthiness judgments function just like other epistemic and non-epistemic values used in theory choice, guiding and influencing scientists' choices in basically the same manner.[14]

[13] This is, notably, significantly broader than the inquiry-ticket approach of DiMarco and Khalifa, who take the object of pursuit-worthiness judgments only to be "a question, where the answer to that question breaks an earlier tie between two competing hypotheses" (DiMarco and Khalifa 2019, p. 1017).

[14] There are interesting questions, to be sure, about the legitimacy or the appropriate role of different kinds of values in different contexts of scientific judgment (considered, e.g., in DiMarco and Khalifa 2019; Shaw 2022). I will not take up those general questions here, since integrative promise involves only one particular set of such value judgments.

One way in which to cast my entire project in this book is, in fact, as an exploration of a particularly prevalent and important kind of pursuit-worthiness judgment in the life sciences. In examining just what epistemic and non-epistemic values function as the end-goal for the pursuit of a particular scientific explanation, I am teasing out commonalities between ways in which life scientists have chosen to fill in the variables *X* and *Z* in Šešelja et al.'s schema. We will see repeated appeals to judgments that certain kinds of explanations are worthy of pursuit precisely because they allow us to advance this particular set of goals, goals which have constituted a central target for centuries of work in biology. But as I have already noted, my argument for the usefulness of integrative promise as an analytic lens won't start from an *a priori* defense of its definition, but rather from a set of applied cases drawn from across the past and present of the life sciences. The first of those cases will arrive in the third chapter, as we consider stringency and opportunism. First, however, we need to delve into one further theoretical interlude, as philosophical reflection on promise's first main characteristic has left us with a host of potential objections.

References

Achinstein P (1993) How to defend a theory without testing it: Niels Bohr and the "logic of pursuit". Midwest Stud Philos 18:90–120

Baker E (2022) From planning to entrepreneurship: on the political economy of scientific pursuit. Stud Hist Philos Sci 92:27–35. https://doi.org/10.1016/j.shpsa.2022.01.013

Bognon-Küss C (2019) Between biology and chemistry in the enlightenment: how nutrition shapes vital organization. Buffon, Bonnet, C.F. Wolff. Hist Philos Life Sci 41(1):11. https://doi.org/10.1007/s40656-019-0248-z

Brown TM (1987) Medicine in the shadow of the *Principia*. J Hist Ideas 48(4):629–648

Brown MJ (2013) Values in science beyond underdetermination and inductive risk. Philos Sci 80(5):829–839. https://doi.org/10.1086/673720

DiMarco M, Khalifa K (2019) Inquiry tickets: values, pursuit, and underdetermination. Philos Sci 86:1016–1028

DiMarco M, Khalifa K (2022) Sins of inquiry: how to criticize scientific pursuits. Stud Hist Philos Science 92:86–96. https://doi.org/10.1016/j.shpsa.2021.12.008

Dobzhansky T (1964) Biology, molecular and organismic. Am Zool 4:443–452

Douglas HE (2009) Reintroducing prediction to explanation. Philos Sci 76(4):444–463. https://doi.org/10.1086/648111

Elliott KC, Richards T (eds) (2017) Exploring inductive risk: case studies of values in science. Oxford University Press, Oxford

Feyerabend P (1975) Against method, 3rd edn. Verso, London

Kellert SH, Longino HE, Waters CK (2006) Scientific pluralism. University of Minnesota Press, Minneapolis

Kitcher P (1981) Explanatory unification. Philos Sci 48(4):507–531

Kuhn TS (1977) The essential tension: selected studies in scientific tradition and change. University of Chicago Press, Chicago

Kuhn TS (1996) The structure of scientific revolutions, 3rd edn. University of Chicago Press, Chicago and London

Laudan L (1971) William Whewell on the consilience of inductions. Monist 55(3):368–391

Laudan L (1977) Progress and its problems: towards a theory of scientific growth. University of California Press, Berkeley

Lichtenstein EI (2021) (Mis)understanding scientific disagreement: success versus pursuit-worthiness in theory choice. Stud Hist Philos Sci 85:166–175. https://doi.org/10.1016/j.shpsa.2020.10.005

Maudlin T (2004) Causation, counterfactuals, and the third factor. In: Collins JD, Hall N, Paul LA (eds) Causation and counterfactuals. The MIT Press, Cambridge and London, pp 419–443

Mayo DG (1996) Error and the growth of experimental knowledge. University of Chicago Press, Chicago

McMullin E (1976) The fertility of theory and the unit for appraisal in science. In: Cohen RS, Feyerabend PK, Wartofsky MW (eds) Essays in memory of Imre Lakatos. Springer Netherlands, Dordrecht, pp 395–432. https://doi.org/10.1007/978-94-010-1451-9_25

McMullin E (1982) Values in science. In: PSA: Proceedings of the biennial meeting of the Philosophy of Science Association, vol 2, pp 3–28

Newton I (1999) The principia. University of California Press, Berkeley

Nicoglou A, Merlin F (2017) Epigenetics: a way to bridge the gap between biological fields. Stud Hist Philos Biol Biomed Sci 66:73–82. https://doi.org/10.1016/j.shpsc.2017.10.002

Potochnik A (2017) Idealization and the aims of science. University of Chicago Press, Chicago

Quinn A (2017) Whewell on classification and consilience. Stud Hist Philos Biol Biomed Sci 64:65–74. https://doi.org/10.1016/j.shpsc.2017.06.007

Russell ES (1917) Form and function. E. P. Dutton, New York

Sánchez-Dorado J (2023) Creativity, pursuit and epistemic tradition. Stud Hist Philos Sci 100:81–89. https://doi.org/10.1016/j.shpsa.2023.05.003

Schindler S (2017) Theoretical fertility McMullin-style. Eur J Philos Sci 7(1):151–173. https://doi.org/10.1007/s13194-016-0150-4

Schindler S (2018) Theoretical virtues in science: uncovering reality through theory. Cambridge University Press, Cambridge

Šešelja D, Kosolosky L, Straßer C (2012) The rationality of scientific reasoning in the context of pursuit: drawing appropriate distinctions. Philosophica 86(4):51–82

Shan Y (2020) Doing integrated history and philosophy of science: a case study of the origin of genetics. Springer, Cham

Shaw J (2022) On the very idea of pursuitworthiness. Stud Hist Philos Sci 91:103–112. https://doi.org/10.1016/j.shpsa.2021.11.016

Soler L, Zwart S, Lynch M, Israel-Jost V (eds) (2014) Science after the practice turn in the philosophy, history, and social studies of science. Routledge, New York

Waddington CH (1959) Evolutionary systems–animal and human. Nature 183(4676):1634–1638. https://doi.org/10.1038/1831634a0

Whewell W (1847) The philosophy of the inductive sciences, founded upon their history, volume II, 2nd edn. John W. Parker, London

Whitt LA (1992) Indices of theory promise. Philos Sci 59:612–634

Chapter 2
Inter-Domain Integration

Abstract In defending integrative promise, I regularly appeal to two senses of integration: a horizontal sense, describing integration across the tree of life, and a vertical sense, describing integration across something like levels of organization. Despite the intuitive appeal of these notions, the idea of vertical integration in particular has been subject to much philosophical criticism. In this chapter, I detail that criticism, in the process building a modified definition of what I will call "inter-domain" integration that, I claim, can capture much of the intuition behind vertical integration without falling prey to the kinds of well-known metaphysical worries to which other forms of vertical integration are subject. Inter-domain integration, by focusing on the ways in which scientists foreground and background different kinds of explanatory resources, more naturally functions as an element of the pattern of explanatory virtues that is integrative promise.

Keywords Integration · Levels · Composition · Organization · Mechanism · Interdisciplinarity

The first feature of integrative promise that I want to explore is its appeal to *integration*. As we saw in the preface and in the last chapter, there is a natural and intuitive sense in which the life sciences employ a structure of "levels" of composition, ranging from molecules to the biosphere, and we see much scientific effort expended in working to *integrate* or *unify* our approaches to those levels.

In this chapter, I want to make that intuition more precise in two important ways. First, this naïve picture of levels of organization, on which we can envision the world as made up of a nesting hierarchy of ever-smaller objects, with each "level" offering its own corresponding science, has been the target of much philosophical criticism—in fact, after nearly seventy years of evaluation in the philosophical literature, it seems to be straightforwardly untenable. How, then, can we save the intuition with which I began my account in the last chapter, which seems, despite its putative incoherence, to be a recurring feature of biologists' approach to the natural world? I'll argue here that we can, at the cost of accepting a fairly loose and pragmatic understanding of what a "level" might look like.

C. H. Pence, *Integrative Promise*, Synthese Library 499,
https://doi.org/10.1007/978-3-031-88471-9_2

However, I claim that a less rigid concept of levels also comes with significant advantages. Most importantly, it will let us understand the drive in biological sciences to what I will call *horizontal* integration (in addition to the *vertical* integration already mentioned): the desire for explanations that can hold across a wide array of *taxa* rather than a wide variety of compositional scales. Both vertical and horizontal integration will be found throughout the examples of integrative promise that we will see in the rest of this book.

And further, combining this approach to levels with an evaluation of the extant literature on the concept of integration in the philosophy of science will allow us to see an illuminating parallel with discussions of *interdisciplinarity*: integrative explanations, as I will make more precise below, are a close cousin of transformative interdisciplinary explanations in science. To conclude, I will introduce a general term for both of these types of integration—*inter-domain integration*—that will be my shorthand for both vertical and horizontal integration as I will use them in what follows.

2.1 Levels of Organization

As we saw in the last chapter, the idea that the world is arranged in something like a hierarchy of levels is familiar from the first few pages of nearly any introductory textbook in biology, ecology, or evolution (see Fig. 1.1 for a representative example). Carl Craver has argued that this only begins to scratch the surface of the apparent diversity of "levels" talk across science and the philosophy of science—comprised of, at the very least, levels of abstraction, aggregation, organization, analysis, causation, explanation, implementation, processing, realization, size, and disciplines (Craver 2015, p. 1). As Jaegwon Kim (2002) has noted, this notion remained largely unstudied for decades, lingering in the background of an array of works in metaphysics and philosophy of science without being carefully analyzed.

For our purposes here, let's zoom in on the concept of levels of organization, which seems to be the most likely candidate for the sense of levels that's at work in these classic cases in the teaching of biology.

There is a traditional presentation of the levels of organization as a kind of guiding idea behind a picture of scientific disciplines, explanations, and reduction that is particularly well encapsulated by the position of the logical positivists. Writing on the "unity of science as a working hypothesis," Paul Oppenheim and Hilary Putnam provide what has become the reference for this framework, now often called the "layer-cake" model of levels of organization. On the layer-cake view, the world can be divided into one, single, univocal system of levels. Each of these levels is studied by a particular branch of science, and the ordering relation between them is determined by the potential for the kind of reductionism that will eventually power the unity of science as a whole:

> a branch B_1 is a *potential micro-reducer* of a branch B_2 if the objects in the universe of discourse of B_2 are wholes which possess a decomposition into proper parts all of which belong to the universe of discourse of B_1. (Oppenheim and Putnam 1958, p. 6)

The language is somewhat complex, but the idea is not: a branch of science (say, physics) is a *potential micro-reducer* of another branch of science (say, chemistry), if it is the case that every object studied by chemistry can be broken down into parts, each of which is studied by physics. The order of levels, then, is just the order of potential micro-reducing disciplines. Fields of science are tightly linked to their level of organization, which is in turn tightly linked to the possibility of a reductive explanation of higher levels in terms of lower levels.

This yields, they argue, a system of levels that is relatively familiar (Oppenheim and Putnam 1958, p. 9):

6. .... Social groups
5. .... (Multicellular) living things
4. .... Cells
3. .... Molecules
2. .... Atoms
1. .... Elementary particles

On the hypothesis of such an overarching system, we can recover the "natural" collection of levels, now recognized to be "'natural' in a deeper sense, of being based on the relation of *potential micro-reducer* obtaining between the branches of science" (Oppenheim and Putnam 1958, p. 28).

This view certainly has merit—the close connection between scientific disciplines, the explanations of systems offered by those disciplines, the levels of organization, and reductionism would mean that many of the independent senses of "levels" picked out by Craver above would actually collapse into a single hierarchy.[1] The trouble is that a single division into layers doesn't seem to work. Perhaps most problematically, it's simply not true that this exclusive partitioning into levels holds for the vast majority of scientific explanations (Potochnik and McGill 2012, pp. 126–127). Living things are not composed *only* of cells—an explanation of the respiration of an organism, for instance, will include not just the behavior of cells in the respiratory system, but will also *directly* include the presence or absence of oxygen molecules. The behavior of social groups, as we have come to learn recently, might be strongly altered by the presence of viruses, some two or three levels lower in the hierarchy.

Further, as a number of authors have argued (e.g., Craver 2015; Kästner 2018; Brooks 2021), the division of a system into levels can also depend crucially upon the epistemic interests of a particular investigator. To take Craver's illuminating example, there might be some sense in which Camillo Golgi (in the late nineteenth century) and Alan Hodgkin (in the mid-twentieth century) were studying "the same level" of phenomena—that is, both were looking at the nature of cells. But the two scientists used different techniques to study different behaviors of different systems; even the conception of 'cell' at work in the two cases would have been radically dissimilar in these two scientific eras. It thus seems to be the case, as Craver notes,

[1] A more recent effort at a unified account of levels, which I lack the space to discuss here, can be found in List (2019).

that even the very parts into which the system could be disassembled would differ, despite their work's superficially describing "the same level" (Craver 2015, p. 7). It doesn't seem here as though we have the kind of timeless, transcendent division into a single set of levels of organization that Oppenheim and Putnam were searching for.

Finally, the idea that each level has its own self-contained forces or causal pathways that largely interact only at that level—an important part of Oppenheim and Putnam's argument that levels can give rise to independent scientific domains—also seems doubtful. As Angela Potochnik and Brian McGill argue, if we think about the ways in which multicellular organisms can interact, those interactions could be mediated by molecules (like pheromones), by cells (like gametes), or by causal processes that reside clearly at the organismic level (like being attacked and eaten; Potochnik and McGill 2012, p. 131). While causal influences may be more *likely* to be separated according to temporal or spatial scales (about which more in a moment), the kind of total separation of those influences that the layer-cake model seems to require is obviously absent.

It is noteworthy that these arguments have a profound tendency to mix ontological and epistemic concerns. After all, the question of the hierarchical composition of the world is a paradigmatically metaphysical one: what, according to our best knowledge, is *really out there* in the world? But this is a metaphysical problem with epistemic consequences. If metaphysics tells us that a given hierarchical structure exists (or doesn't exist)—and especially if that structure comes with restrictions on the causal relations that can be said to obtain within that level or between levels, as we just saw above—then the metaphysical picture is going to constrain what kinds of explanations or theories we are able to produce and defend. Likewise, conversely, if something like the project of naturalized metaphysics is in play, and we thus take our best scientific theories as a guide to our metaphysics, then changes at the epistemic level will have direct consequences for metaphysics.[2] On a different understanding of the relationship between metaphysics and epistemology, one might divorce these two questions, arguing that while the world could (truly, metaphysically) be divided into hierarchical levels, it could (scientifically, epistemically) remain the case that such a division is not useful, or vice versa.[3] Pursuing the consequences of this would take me too far afield here, and run afoul of my own predilections toward naturalized metaphysics.

The failure of this traditional approach to levels of organization has led some to argue that the entire concept of levels should be eliminated from scientific theorizing (Thalos 2013), or, with a bit more nuance, that talk of levels can be replaced without loss by other, more important and more fundamental concepts, such as composition or scale (Eronen 2015). A genuine eliminativism about levels in practice, however, even were an airtight philosophical argument for it available, is likely an impossible goal. Craver has it exactly right when he notes that

[2] For more on the idea of and prospects for naturalized metaphysics in biology, see, e.g., Slater (2017) and Triviño (2022).

[3] Thanks to an anonymous reviewer for offering this potential response.

> The suggestion that we might be better off abandoning the levels metaphor is about as likely to win converts as the suggestion that we should abandon metaphors involving weight or spatial inclusion. These metaphors are too basic to how we organize the world to seriously recommend that they could or should be stricken from thought and expression. (Craver 2015, p. 2)

That said, it is evident that the classic model of Oppenheim and Putnam (1958) is too simplistic to help us understand just what is going on in contemporary presentations of levels of organization. A number of authors have offered alternative accounts, designed to capture something about what has made the levels metaphor so attractive while avoiding its pitfalls; let's examine a few of those efforts.

2.1.1 Levels of Mechanisms

One of the most influential proposals to replace the classic account of levels of organization comes from the robust contemporary literature on the concept of mechanism in the life sciences.

To see how this account of levels might work, consider a now-standard depiction of a mechanistic system, as found in Fig. 2.1. One of the clearest benefits of a mechanistic account, as has been argued since the inception of the mechanistic view (Machamer et al. 2000), is the fact that mechanisms can be decomposed into parts,

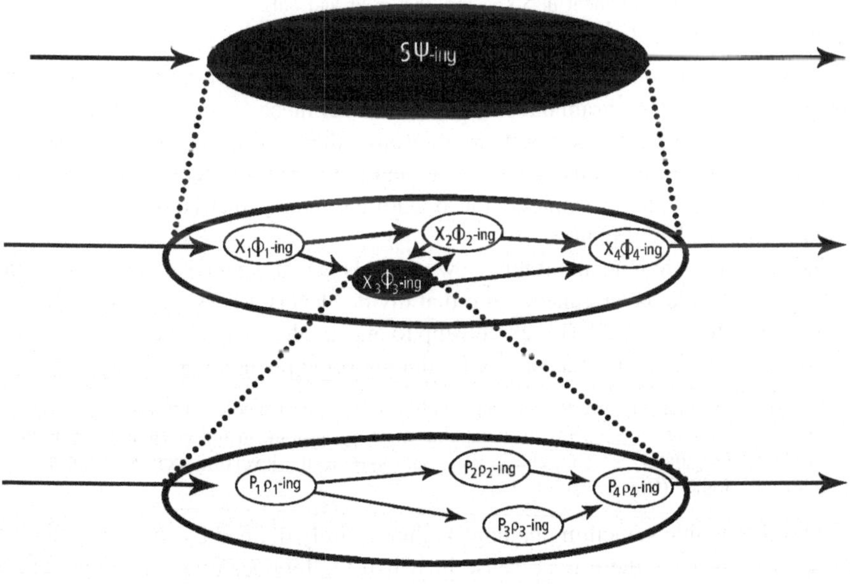

Fig. 2.1 A diagram of a mechanism, demonstrating the ways in which mechanisms can be decomposed into parts which are themselves, in turn, mechanisms. Figure 4 reprinted from Craver (2015, p. 17), by permission of the author and publisher

which in turn might themselves function as mechanisms. In the terms of the figure, we begin by analyzing the overall activity of S ψ-ing into its component parts. This overall activity can be expressed in terms of the interaction of four parts, X_1 to X_4, engaging in their own activities, ϕ_1 to ϕ_4. But we can then focus our attention just on one of those activities of one of the parts: here, X_3 ϕ_3-ing. That activity, in turn, can be decomposed into a mechanism, here the interaction of four parts P_1 to P_4 engaged in their own activities ρ_1 to ρ_4.

Let's make this more concrete, following Craver, who is probably the leading advocate of such a mechanistic picture of levels (Craver 2015, p. 17; see also Craver and Darden 2013, pp. 161–185). The highest-level activity here might be a rat's learning to run through a maze (thus the rat is S, and ψ is a spatial memory task). At the next level down, one of the parts (X_3) is the hippocampus, which is engaged in forming a "map" of the local spatial relations around the rat (ϕ_3). That process, in turn, is known to involve a process called Long-Term Potentiation (LTP, say, ρ_2) in synapses between pyramidal cells (P_2). (We can go deeper, adding another level to look at the role of one particular kind of receptor on the relevant cells, or yet another phrased in terms of the actual conformal changes occurring in proteins.)

That there is a kind of nested hierarchy of levels at work in cases like these is probably already intuitively obvious (or at least is obvious once we see Fig. 2.1). But we can make the idea precise, and making it precise is important for seeing both the advantages and the limitations of this view. What we need is a definition of the idea the parts P are "at a lower level" than the parts X:

> Some component, X's ϕ-ing, is at a lower mechanistic level than S's ψ-ing if and only if X's ϕ-ing is a component in S's ψ-ing, that is, if and only if X's ϕ-ing is a relevant spatiotemporal part of S's ψ-ing. (Craver 2015, p. 17)

Thus, returning to our example, both the hippocampus's forming a local map and the process of LTP in pyramidal cells are at a lower mechanistic level than the rat's spatial memory task, because both are relevant—that is, both actually contribute to the performance of the activity by the rat, and both are spatiotemporal parts of the actual mechanism that permits the rat to learn the maze—that is, both occur in the brain during the learning task.

The first thing to note about this view of levels is that we have come quite a long way from a single, layer-cake picture that divides levels across all of the sciences. Indeed, two items have a level relationship to one another *only if* they appear within the same overarching mechanism for producing some phenomenon.

> Levels of mechanisms are defined only within a given part-whole hierarchy. There are different levels of mechanisms in the spatial memory system, in the circulatory system, in the osmoregulatory system, and in the visual system; the levels in each need not map onto one another. (Craver 2015, p. 18)

The level to which something belongs is thus entirely defined by the mechanism at issue—for instance, there is no sense in which the hippocampus, a component of the brain, is at the "same level" as the epidermis, a component of the skin. This also entails that there is no unique answer to "the level" of a particular token component: the level of a sodium ion, for instance, might differ dramatically across different neural mechanisms depending on how "deep" one has to travel to arrive at that ion in the description of the mechanism under discussion at the moment.

Perhaps more problematically, this means that some questions about whether objects are at the same or different levels that we might intuitively expect to be answerable are in fact not, according to Craver's view. Imagine, for instance, that in Fig. 2.1, in addition to the decomposition of X_3 into P_i, we decompose another part of the brain, X_2, into its parts Q_i as well. Now, what happens if one of those Q_i involves exactly the same entities performing exactly the same activities as one of the P_i? In terms of Craver's example in rats, imagine that, in addition to LTP taking place in the hippocampus (X_3), there is also LTP taking place in X_2 as well. P_2 and Q_2 are, then, both synapses between a certain kind of neuron, and ρ_2 and χ_2 (the activities which those parts perform) are both the process of long-term potentiation. Both P_2 and Q_2 are certainly, following Craver's definition, *at a lower mechanistic level* than the rat's learning the maze. But it is not clear that there are resources in this theory to capture the intuition that the process of LTP should be at the *same level* when it features in these two sub-mechanisms.[4]

Markus Eronen (2015) distills these various objections into two broad-scale problems for the mechanistic view of levels. The first is that it's not clear that invocations of "levels" in scientific literature (like "the molecular level" or "the population level") indeed refer to levels of mechanism. This captures something of the worry just discussed: when neurophysiologists refer to the "cellular-electrophysiological level," one might expect that they are making reference to things like LTP, though as we just saw, various instances of LTP aren't in fact guaranteed to be found at the same level on the mechanistic account (Eronen 2015, pp. 44–45). More broadly, it seems that levels talk like that we cited in textbooks is describing something more similar to the original idea of Oppenheim and Putnam (1958), rather than the localized sense of levels of mechanisms (Eronen 2015, pp. 46–47). It's thus unclear in general whether a mechanistic definition of levels is capturing the phenomena to which scientists intend to refer when they discuss levels.

Eronen's second major worry has to do with the notion of "component" that appears in Craver's definition of mechanistic levels. The definition that I cited above involves an extremely weak notion of component—anything that is a "relevant spatiotemporal part" can qualify as a sub-component of a mechanism. But in that case, strictly speaking, *all* of the X_i as well as *all* of the P_i should qualify as "components" of the mechanism for S's ψ-ing, as they are all relevant spatiotemporal parts. That leads to the strange conclusion that X_3 and P_1 are both components of S (thus, in a sense, "at the same level") and, at the same time, that P_1 is a component of X_3 (thus, in another sense, "at different levels"). This seems contradictory and, Eronen argues, should lead us to adopt a notion of *direct*

[4] There is an intuitive response available in this particular case: that we had to "descend" the same number of steps to move from the overall memory task to the activity of LTP. But this is a contingent feature that won't generalize. P_2 could well instead be a sub-sub-sub-mechanism of X_2—a change that we might make only by altering the way in which we choose to break the mechanism down into its components, not by any difference in the empirical facts of the matter.

components instead: only the X_i are direct components of the overall S-mechanism, and only the P_i are direct components of X_3. But in this case, the definition of "membership" at each level becomes so narrow that each level has very few direct components (Eronen 2015, pp. 49–50).

It thus seems apparent that a mechanistic account of levels is, as Craver has argued, a way to give the notion of levels "a relatively precise and metaphysically unobjectionable formulation" (Craver 2015, p. 23), and one which tracks some uses of levels, especially in molecular biology. But it is less clear that the structure of levels that results really tracks the notion of levels of organization that motivates my project here.

This seems all the more true when we think about the possibility of using levels of mechanism to ground an account of vertical integration. If we can express a mechanistic diagram like Fig. 2.1, then we *already have* a vertically integrative explanation that can bring together S, the X_i, the P_i, and their activities—especially if to offer a mechanistic explanation *just is* "showing or demonstrating that the phenomenon to be explained is a product of one or more of these abstract and recurring types of activity or the result of higher-level productive activities" (Machamer et al. 2000, p. 22), that is, if giving a mechanistic explanation *just is* demonstrating the mechanism responsible for the phenomenon. While some think that the demonstration of mechanisms in this way is exhaustive of the concept of scientific (or, perhaps better, causal) explanation, many others disagree. If there is indeed *no* mechanism in which some objects participate, then things get even worse for our putative account of integration. On the levels-of-mechanism account, these objects would simply sit outside the hierarchy of levels entirely, and we wouldn't be able to justifiably say whether or not they were on the same or different levels—the question would not even make sense. It therefore seems that levels of mechanism are unable to ground an account of vertical integration.

2.1.2 Levels and Scale

The levels of organization that we've already discussed (and that biologists seem interested in integrating) are therefore not directly representable in terms of levels of mechanisms. An alternative account might focus on another common feature that separates them: *scale*. Atoms, molecules, cells, organisms, and ecosystems have both characteristic spatial sizes and characteristic temporal scales on which their interactions take place. This was taken to be central to accounts of levels by William Wimsatt (1974, 1976), who argued that levels designate scales at which causal interactions are particularly comprehensible and repeatable. In the years since, a number of authors have defended accounts of levels as centered around concepts of either spatial or temporal scale (Potochnik and McGill 2012; Eronen 2015; DiFrisco 2017).

To put the basic insight differently, as Eronen has noted, we can learn something from the way in which the mechanistic account failed to give us a clear understanding of the "same-level" relationship. If we want to "make the same-level relation plausible, something more than just compositional relations seems to be required, such as similarity of scale or causal interactions" (Eronen 2015, p. 51). What makes the "molecular level" molecular or the "population level" populational cannot only be its components or the collectives that it constitutes—since, as we've seen, those might have too diverse a nature to pick out a single level—but must be some combination of compositional facts with other facts. The first such candidate we'll examine for those further facts is the notion of scale.

The most general account of scale comes from Potochnik and McGill. For them, drawing on the ecologist Simon Levin, "scale is the spatial or temporal extent across which observations span" (Potochnik and McGill 2012, p. 133). Ecological papers have, for some time, delineated the scales at which their claims are expected to hold—as they put it, ecology has "become explicitly scale dependent" (Potochnik and McGill 2012, p. 134). Since different theories and different interventions tend to succeed at different levels, Potochnik and McGill argue that this gives us all we need to construct a scheme of "quasi levels" that associates the relevant spatial scale with the theories, explanations, or techniques applied to it; "cells occur on a scale of microns to millimeters, while tissues are typically millimeters to decimeters; organisms (vertebrate and angiosperm, at least) are centimeters to many meters; ecosystems are measured in kilometers; and the biosphere is much larger still" (Potochnik and McGill 2012, p. 135).

The most significant objection to such a view, and the reason for which the authors call these "quasi levels," is the fact that since scale is a continuous variable, there's no single, discrete ordering of scales into which the biological world might fit. Unlike in the case of the layer-cake view, there's no sense in which "being on a different level" can be specified independently of an investigator and their theoretical and empirical context: sometimes an individual squirrel and a tree might be of the "same" scale (e.g., when talking about seed dispersal), and sometimes they might not (e.g., when talking about the development of masting behavior, which requires a tree to interact with a *population* of squirrels). And we don't necessarily care about part-whole relations at all anymore—while these may sometimes be relevant, levels in this sense are doing more work in *tracing causal interactions* than picking out branching structures of composition. We'll come back to these questions of the epistemic utility of the levels concept in the next subsection.

In defending a similar approach,[5] James DiFrisco tackles this objection head-on, arguing that rather than dismissing this scale-based structure as "quasi levels," the concept of levels provided by *temporal scales of causal interaction* is the best for which we might hope. As he writes,

> Rather than a universal and discrete division of causal interactions or types, to the contrary, it is both more plausible and closer to the scientific use of the concept to think that

[5] DiFrisco has motivated reasons for appealing to the temporal scale of causal interactions rather than their spatial scale; unfortunately I lack the space here to litigate the debate between the view of DiFrisco and that of Potochnik and McGill.

> a hierarchy offers *reliable generalizations* about the structure of causal interactions in complex systems—not universally for any and every system, and not without some minor exceptions to strict discreteness. (DiFrisco 2017, p. 801)

Thus even though these levels will not be universal (they will not cut across the entirety of the world in exactly the same manner), and will not be fully discrete (as cases like those of the squirrels and trees will occasionally blend levels together for pragmatic purposes), we still have a significant and reasonable division of the world into levels, one which will become "more informative the more general and discrete they are" (DiFrisco 2017, p. 800).

The scale-based approach to levels thus constitutes one of the more promising "deflationary" views of levels, one that structures our concept of levels around a feature that is undeniably an essential part of what we had in mind in proposing the concept in the first place. Before summing up and turning to what these theories of levels might mean for my notion of vertical integration, I want to look at one more important family of approaches to levels.

2.1.3 *Epistemic Conceptions of Levels*

One insight present in both Craver's and DiFrisco's approaches to levels can give us another way in which to understand how this term is used. For both of these authors, we should consider levels not merely as an ontological or metaphysical feature of the world, but further—and perhaps more importantly—in terms of the epistemic goals that any concept of levels is intended to satisfy. Eronen reconstructs a handful of such goals: providing a framework for understanding reduction; capturing significant features of the organization of nature; clarifying talk of levels in the sciences; and providing a framework for analyzing downward causation (Eronen 2015, p. 53). Craver's view, Eronen argues there, seems to target all four of these goals. DiFrisco, as we just saw, explicitly frames his discussion in terms of the ability of a levels account to extract reliable generalizations about causal behavior.

A treatment of levels that makes these epistemic goals central has been significantly expanded in recent years by a number of additional authors. Daniel S. Brooks has argued that levels refer to "a number of ontological and epistemic things simultaneously…, which are further contextualized to specific claims made by the authors in the text" (Brooks 2021, pp. 1492–1493). Again, what distinguishes those things are the epistemic goals for which they are deployed: here, "partitioning a problem space," "coordination within the problem space," and "locating within a problem space" (Brooks 2021, pp. 1494–1497).[6]

A bit more broadly still, Lena Kästner, drawing on other work, deploys her concept of "epistemic perspectives," which combine differences in resolution, speci-

[6] This expands on an earlier view presented in Brooks and Eronen (2018), where levels are presented as a kind of "heuristic."

ficity, point of view, sensitivity, and scope, to craft a theory of levels grounded in this more expansive understanding of scientists' epistemic resources. Unfortunately for our purposes here, she argues that "two entities *A* and *B* are 'at the same level' … if and only if *A* and *B* can both be observed from the same epistemic perspective" (Kästner 2018, p. 74). Vertical integration, however, seems to require the ability to reidentify the same entity across different perspectives—integrating the work in one perspective on one entity with work *at the same level* arising from a different perspective. But just as we saw with Craver's view, it's not clear that Kästner's theory allows us to perform this kind of reidentification across locally-contextualized notions of levels that vertical integration would require.

The most promising epistemic approach has recently been published by Eric Hochstein (2022). For him, levels of organization arise relatively organically from a process of *foregrounding* and *backgrounding* that scientists are forced to make when they construct explanations. In his words,

> levels are best understood as choices that working scientist[s] make regarding how and why particular kinds of information are being foregrounded in their representations, while other kinds of information are pushed into the [background]. To change levels is to change the foregrounding and backgrounding of information in our representations for different representational and pragmatic purposes. (Hochstein 2022, p. 2)

To be sure, this account is subject to the same kinds of worries about lack of discreteness and continuity that we saw above for scale-based views of levels. But in that it explicitly underlines the fact that foregrounding and backgrounding are deliberate choices, made for both epistemic and non-epistemic reasons, it presents a theory of levels of organization in line with the general approach to explanatory virtues that I will pursue here.

There are also promising solutions available to such concerns about discreteness. Since foregrounding and backgrounding are also, at least in part, matters of scientific communication set by disciplinary standards, the structure of scientific communication will in turn impose at least some measure of discreteness on the levels that Hochstein's theory identifies. As he puts it, "whether two descriptions operate on the same, or different, levels often depends on what the models are being used to represent, and what the model-builder thinks that the consumer of the representation expects" (Hochstein 2022, p. 13). This makes an obvious place for us to anchor a concept of integration: vertical integration would simply be the construction of models that foreground a collection of information that model-consumers wouldn't expect to be simultaneously foregrounded, given their extant disciplinary commitments.

This concept of levels also seems to be able to take on the insights of the scale-based view while explaining away some of its defects. Emphasizing the importance of disciplinary standards and explanatory virtues gives us a motivated and justifiable reason for the "fuzziness" of the hierarchy of levels: the level structure will be fuzzy just insofar as the constellation of epistemic and practical goals deployed within a particular branch of the sciences are themselves fuzzy. Such fuzziness becomes, on this view, an inescapable feature, but one which we are now given tools to understand and analyze. We can also see a way to integrate DiFrisco's understanding of levels as offering reliable causal generalizations. It is entirely unsurprising that these processes of foregrounding and backgrounding would in no small part be motivated by the desire to ground reliable causal generalizations—this is quite plausibly the paramount virtue in the construction of scientific models, and so its appearance in an account of levels parallels its importance for the scientific disciplines involved.

Moreover, and finally, we can see an opening for precisely the kind of two-dimensional account of vertical and horizontal integration that I want to develop here. For there is no reason to think, *a priori*, that these processes of foregrounding and backgrounding as described by Hochstein will only result in a vertical account of levels, roughly corresponding to the compositional hierarchy. Precisely the same kind of process in model construction seems to inform the distinctions that separate microbiologists from botanists, primatologists from ornithologists. In short, we have here a tool that can potentially explain both vertical and horizontal integration. Before making that story explicit, however, we should clarify another aspect of our terminology: the notion of integration itself.

2.2 Integration and Interdisciplinarity

The literature on explanatory integration in philosophy of biology and general philosophy of science is extremely diverse. Not infrequently, integration is discussed within the context of particular sciences, or in the elaboration of theoretical frameworks surrounding knowledge transfer.[7] Theoretical considerations of what explanatory integration is and how it might work also abound. While it would be folly to attempt a complete survey of this material here, what I want to do in this section is extract a moral that cuts across this work, one that will allow us to bring together discussions of integration and on interdisciplinarity. In short, whatever further detail we might later add to the story, explanatory integration functions as a kind of transformative interdisciplinarity.

To begin, consider Lindley Darden and Nancy Maull's classic work on interfield theories (Darden and Maull 1977). A *field*, in their sense, consists of a problem, a domain of relevant facts, a set of goals or norms, a set of techniques and methods for attaining those norms, and potentially a collection of concepts, theories, laws, or terms. Interfield theories, then, arise when we need to answer "interfield problems," or when we need to explain the relationships between multiple fields.

[7] Examples include Ingo Brigandt's discussions of evolutionary novelty and systems biology (Brigandt 2012, 2013); Anya Plutynski's discussion of "integrative science" in the case of cancer biology (Plutynski 2013); the importance of integration for what Adrian Currie has called "comparative thinking" in evolution (Currie 2019, 2021); or, from a very different perspective, Rens Bod and collaborators' discussions of integration in the framework of the "flow of cognitive goods" in the history of science (Bod et al. 2019).

They enumerate a number of examples of what such interfield constructions might look like: one field's spatially locating an entity or process from another field, describing its physical nature, or its structure, would all count as interfield theories, as would describing causal connections between fields (Darden and Maull 1977, p. 49).

More interesting for my purposes is their summary of the *function* of interfield theories. In addition to "solving," in some sense, the interfield problem that led to its development, a true interfield theory must both answer questions which "cannot be answered using the concepts and theories of [one] field alone," and possess a kind of novelty, either drawing attention to "previously neglected items of the domains of one or both fields," predicting "new items for the domains" of the fields, or launching "new lines of research" (Darden and Maull 1977, p. 59). In short, a genuine interfield theory should not only conjoin explanations, methodologies, techniques, and so forth from two different fields; it should result at least to some degree in a genuine change in the way in which those fields operate. It is the extent and nature of this kind of transformation to which I will return below.

Before I do, however, we should investigate the comprehensive work of Ingo Brigandt on explanatory integration in science, perhaps the closest effort in the literature to my project here. Brigandt is largely preoccupied with questions of fundamentality, arguing that "which theoretical approach is explanatorily more fundamental" in a particular case of integration "*varies with the specific problem* at hand" (Brigandt 2010, p. 297, original emphasis). Also setting aside the reliance on mechanisms to privilege a case study of non-mechanistic integrative explanations in evolutionary developmental biology, his central example is the concept of evolutionary novelty or innovation.

What makes this case interesting for integration as a whole, he argues, is the fact that the kind of integration it involves fails to give a "*stable* [set of] theoretical and disciplinary relations across several scientific disciplines, where each such discipline has a *unique* set of relations to others" (Brigandt 2010, p. 306). If such stable and unique relations were present, we might be able to talk about unification or synthesis, rather than integration. Integration, by contrast, is marked by the fact that explanatory integrations might hold only in the context of one particular scientific problem, and hence integration involves "more *transient* relations to other fields, depending on which problem is currently addressed" (Brigandt 2010, p. 306).

So far so good. If integration is, as I argue here, an explanatory virtue, then (as I will discuss further below) it is "problem-centered" in just the sense that it is to be evaluated in the context of a particular scientific explanation. Brigandt is thus correct that integration will be thereby limited in scope. The fact that integrative promise also applies to the pursuit of *future* explanations can, however, temper this "transience" somewhat: our ability to envision the promise of additional lines of integrative research can give the notion a future-directedness that the synchronic account of Brigandt lacks.

There is a more significant problem for my account found in Brigandt's work, however. For he goes on to argue that,

> from the present perspective, it is not the case that unification/integration is an a priori condition that is to be achieved in every context. Integration is not an aim in itself, but integration is likely to be needed *for the aim of* solving a complex scientific problem. [. . .] To which extent integration and specialization are needed is not an a priori matter but depends on the case. (Brigandt 2010, p. 307)

We could certainly read Brigandt as arguing here that integration in fact does not serve as an explanatory virtue—that integration is not "an aim in itself." But let's unpack the argument in a bit more detail. Integration, he claims, is valued only because it is instrumentally useful for solving a particular scientific problem, and therefore it cannot be an explanatory virtue in general, but only in particular cases. I want to respond to this argument in two ways.

First, seeing integration as an explanatory virtue does not mean that it is "to be achieved in every context," nor that it is "an a priori matter" that integration will be pursued by every scientist in the process of explaining a phenomenon. This kind of subject-relativity is already well known from other literature on theoretical virtues as discussed in the last chapter: a virtue like simplicity might only be pursued in very particular contexts, by very particular scientists. Put differently, the fact that integration is only valued in some circumstances does not entail that it is not a virtue.

Second, I think we will see across a number of the case studies I canvass in this volume that, indeed, scientists *do* often pursue integration as "an aim in itself"—or, at the very least, they cite integration as one of the reasons that the explanations that they offer should be valued and accepted by the scientific community. The weight of that argument will have to be assessed over the remainder of the book.

Setting aside this disagreement, then, Brigandt's account of integration is extremely amenable to my approach. Even more in line with my use of integration is his perspective when Brigandt returns to the notion a few years later. He argues there that *problem agendas* (sensu Love 2008) "coordinate interdisciplinary research" by specifying "a particular *epistemic aim*," which in turn sets "associated standards of explanatory adequacy" (Brigandt 2012, p. 83). While I don't want to take on Love's entire problem-agenda framework here, I think that seeing integrative explanations as *concretely exemplifying* standards of explanatory adequacy—or explanatory virtues—in cases of integrative promise is a very similar depiction of integration to the one that I am offering.

After a long summary of these and other accounts of explanatory integration on offer in the literature, Anya Plutynski sums them up by saying that nearly every author in the debate is looking not only for the mere existence of shared concepts, data, or methods, but, further, at least "(a) a transformation in researchers' conception of the object of explanation, or (b) a change in what would count as an adequate explanation of that object, or (c) both" (Plutynski 2013, p. 468). This parallels the presentation of Brigandt and Darden and Maull above—and for anyone acquainted with the literature on interdisciplinarity, this kind of claim will sound extremely familiar.[8]

Discerning how we can separate cases of true, transformative interdisciplinarity from more simplistic and limited cases has long been a preoccupation of the literature on interdisciplinary investigation. While a full review of that work would take me much too far afield, we can draw some interesting insights by considering the taxonomy of interdisciplinarity offered by Julie Klein (2010).[9] While she subdivides interdisciplinary research in more ways than would be useful for us here, something of what is distinctive about the interaction that Plutynski notes is important for cases of integration. We first set aside "multidisciplinarity," as a kind of encyclopedic gathering of facts from various fields without any significant interaction. Turning to interdisciplinarity itself, after climbing a loosely ordered hierarchy of meanings of the term, she explicitly discusses integration at the very last level, what she calls, following Margaret Boden, "integrated interdisciplinarity":

> In Integrated ID ... the concepts and insights of one discipline contribute to the problems and theories of another, manifested in computational neuroscience and the philosophy of cognitive science. Individuals may find their original disciplinary methods and theoretical concepts modified as a result of cooperation, fostering new conceptual categories and methodological unification. (Klein 2010, p. 20)

Perhaps unsurprisingly for a view calling itself "Integrated ID," this comes very close to the common thread that Plutynski noted tying together the varied accounts of integration in the literature. We want not only to aggregate, but to reshape, to reconstruct the fields that are integrated. Arguably, this is part of what makes integration valuable in the first place.

To sum up, then, as I will use the term here, *integration* is the deployment of theories, concepts, methods, and techniques from multiple domains of science in the context of a single explanation, where this deployment transforms the explanations that would be provided by any one domain alone.[10] Integration is related to both theoretical unification and to interdisciplinarity, but insofar as I am limiting my definition to integration's role as an explanatory virtue, it is at best a special case of each.

Two more points should be clarified before we turn to the particular examples of vertical and horizontal integration in my sense. First is a potential objection to the importance of integration, raised in the context of idealization by Potochnik. As she argues, scientists often have good reason to preserve some idealizations in cases

---

[8] The same connection has been noted by Celso Neto, who writes that one sense of integration common in the biological literature is the recognition by two communities of biologists "that their own research questions fit into a broad interdisciplinarity" (Neto 2020, p. 7). While he also mentions another sense of integration—the integration of explanatory practices—the two surely lie on a continuum, and every integration will bear at least some of the hallmarks of this recognition of an interdisciplinary dimension to research.

[9] Other helpful sources include Mayntz (1992), Schmidt (2008), and Huutoniemi et al. (2010). I will return to a more practical and normative discussion of interdisciplinarity in Sect. 9.1.

[10] I will return to the concept of domains in Sect. 2.4.

where disciplines could otherwise be integrated, keeping field-specific, separate explanations of phenomena even when those same scientists might themselves also pursue integrative work. Put differently, the pursuit of integration need not entail that the resulting, integrated product *supersedes* the explanations that are thus integrated:

> My point is that even integrative approaches motivated by the shortcomings of existing research projects do not supersede their predecessors by providing a single, integrated treatment of the focal phenomena. Instead, new approaches supplement existing research by representing a causal pattern that has not yet been adequately captured. (Potochnik 2017, p. 77)

My response to this objection is qualified acceptance. That is, I think Potochnik has it right that, at least sometimes, scientists don't have goals to replace or supersede on their minds when they support an explanation for its integrative promise. At the same time, we'll see in some of the cases that I discuss in the rest of the book (especially the example of natural history in Chap. 5) that sometimes promise—or lack thereof—is seen to be an existential question for the discipline at issue. I take this to be a contingent, empirical matter, that can (and does) come out either way.

Second and finally, exactly what is it that is being integrated in these cases, and exactly what is it that is doing the integration? Plutynski notes that integration is often described as a kind of "meta-methodology" rather than a property of something like theories or explanations (Plutynski 2013, p. 469). At the same time, Darden and Maull clearly described integration as being performed by interfield theories themselves. Brigandt, casting his study of integration in terms of epistemic standards and aims, focuses on the role of concepts as the vehicles of integration. Consonant with the discussion of explanatory virtues and integrative promise in the last chapter, I want to argue that integration (at least in the sense that I have described here) is an explanatory virtue, that is, a good-making feature of particular scientific explanations.

In part, this slipperiness surrounding the nature and target of integration arises from its relationship to the history of the philosophy of science, particularly to debates over reductionism. Integration is taken to be importantly distinct from unification; where unification is targeted at the reduction of one theory to another theory with which it has been unified, integration is supposed to be, instead, both "problem-oriented" and "compatible with a modest pluralism" about science as a whole (Plutynski 2013, p. 469).[11]

I agree with the overall orientation of this argument—namely, that integration is, indeed, often problem-oriented, and also is compatible with scientific pluralism. As we will see in several of the examples that I will discuss in the rest of the book, integration is even sought by scientists who wish to defend a kind of pluralism against a threat of reduction. Indeed, an explanation can be integrative in using

[11] As noted in Chap. 1, this tension between integration and unification is part of the reason I wanted to avoid a tight link between integrative promise and theoretical unification. At least in the sense of, for instance, Philip Kitcher's classic view (1981), unification can certainly be read as working in the service of reduction.

concepts and tools from multiple theories, interacting in such a way that it is impossible for one to successfully be reduced to the other (the example of planetary biology in Chap. 4 will be a case of just this type); as Darden and Maull put it for their case of interfield theories, "one field does not reduce another field; reduction in the sense of derivation would be impossible between such elements of a field as techniques and explanatory goals" (Darden and Maull 1977, p. 48). Treating horizontally integrative explanations as potentially reductive is absurd; no one would argue that an evolutionary explanation that integrated the study of bacteria and plants was somehow reducing botany to microbiology, or vice versa.

But there is no reason to think that the twin questions of what is doing the integrating and what is being integrated are directly tied to this issue of reductionism. Put differently, we could defend that it is theories, or explanations, or problem agendas that are playing the key role in integrating two scientific theories, without necessarily committing ourselves to the acceptance or rejection of a thesis about reductionism more generally. A theory- and belief-centric view of science certainly may have led both to, as Brigandt (2012, pp. 75–76) has argued, a neglect of epistemic aims and standards, and to, as Darden and Maull (1977, pp. 59–61) have argued, an overemphasis on reduction. But these two sets of questions can, and should, be disentangled, even if they may have had a historical common cause.

To that end, I claim that what is *doing the integrating* in many concrete cases are scientific explanations, and hence that integration is an explanatory virtue—a property of scientific explanations that guides scientists in choosing which explanations to pursue and to value. After all, we evaluate the success or failure of interdisciplinary work where the rubber meets the road: is it or is it not the case that a specific explanation does, in fact, integrate two disparate domains of scientific investigation? Should we admit a more speculative explanation, or one that is supported by a smaller amount of empirical data, if it holds out the promise of this kind of integration? These are questions that we ask in particular explanatory contexts: integration serves, in these kinds of evaluations, as an explanatory virtue.[12]

So much for what is doing the integrating: my answer is "scientific explanations." What, then, is being integrated? As I have already mentioned, there are two especially pertinent ways to answer this question in the context of the contemporary life sciences, and, finally, a way in which we can generalize over them both.

2.3 Vertical and Horizontal Integration

When explanations in the life sciences do make appeals to integration, what is it that they might be integrating? In part, the best way to answer this question is to

[12] This nicely echoes Ingo Brigandt's claim that integration is a question of epistemic aims and standards (Brigandt 2012, pp. 82–83), which seem to have a similar function as norms for explanation.

examine the suite of examples that make up the bulk of this book. But at a general level, we can note two common trends. The first, *vertical integration*, relates to the question of levels of organization as discussed above. A vertical integration is one which integrates resources found at different scales or compositional levels—for instance, an explanation that combines biochemical and cellular information in the service of an explanation of evolutionary change. A horizontal integration, by contrast, is one which combines resources normally used in distinct areas of the tree of life—for instance, an explanation that combines the epistemic or explanatory resources of botany and microbiology to offer a single account of a shared biological phenomenon.

On the one hand, I think it is intuitive enough that there are two different "axes" of integration at work here, each playing on a different sense in which a biological explanation could "integrate" multiple domains of the life sciences. Biologists are interested, as we will see throughout what follows, in explanations that can tie together the stories we tell at different scales and in different clades.[13]

On the other hand, the account of levels that I have sketched in the first half of this chapter seems precisely to indicate that the picture of levels at work in vertical integration is untenable. Recognizing this difficulty, then, forces me to make two important qualifications to this intuitive account. First, while I will continue to speak of vertical integration, it must be kept in mind that the use of "vertical" here is metaphorical at best. Vertical integration does not imply, nor does it need, a strictly nested compositional hierarchy. Rather, following Hochstein (2022), it requires only that there are different patterns of foregrounding and backgrounding at play in the kinds of explanations that are typically deployed by, for instance, biochemists and evolutionary biologists. That these different patterns exist is, I take it, entirely uncontroversial, and so this less rigid concept of "vertical" integration still makes sense.

Second, and more importantly, the failure of the strict version of the levels metaphor in fact offers me a unified foundation for vertical and horizontal integration. For if what matters is the patterns of foregrounding and backgrounding at work in different domains, then just as these patterns will differ between cellular biologists and ecologists, they will also differ between botanists and ichthyologists. What counts is the way in which these explanations are constructed—the processes of selecting explanatory resources, establishing of explanatory norms, and so forth—and the kind of explanations that propose to reach beyond the standard limits of those norms for some particular domain.

Throughout what follows, I will still use the terms vertical and horizontal integration, to refer to these large axes of integration-across-scale and integration-across-taxa. But a particularly liberating aspect of abandoning a rigid division of the world into levels is that we no longer have to decide whether a given integration is an

[13] These same two axes of integration, for instance, are the first two dimensions of integration described by Sabina Leonelli (2016, p. 143). She also adds a concept of "translational integration," which is outside of my scope here.

instance of one or the other. Consider, for example, an explanation that proposes to integrate accounts of a phenomenon offered in yeast and in mammalian physiology. There is an important distinction in size scale between the entire organisms here, but if the mammalian explanation relies largely on cellular details, it won't be clear whether that difference in size scale actually matters. Similarly, there is, of course, a large taxonomic difference between fungi and mammals. But one could also imagine that the explanation relies on conserved proteins that make this vast taxonomic distance less relevant than it might first appear. Is this integration, then, vertical or horizontal? There may not be a good answer, or the answer may depend on which fine-grained details of the case we think are important.

These corner cases, however, need not trouble our account. All we need in order to say that this is a genuine integration is the fact that the explanations deployed by yeast biologists and those deployed by mammalian physiologists usually foreground and background significantly different facts, methodologies, and concepts, and that the proposed integrative explanation draws from them both while significantly transforming those aspects of the integrated fields in the process. Vertical and horizontal integration do not have to constitute a precise, exhaustive dichotomy.

2.4 Inter-Domain Integration and Promise

It is for this reason that I have chosen to refer to the general concept of integration, whether vertical or horizontal, as *inter-domain integration*. I've selected *domain* in an effort to be as neutral as possible about the scientific community structure that is invoked when we talk about inter-domain integrations. What matters for my purposes is that domains propose scientific explanations, and thus can be the kinds of things that can feature in integrative explanations as I described them above.

In introducing the notion of foregrounding and backgrounding, Hochstein considers both epistemic, representational aspects as well as practical aspects such as methodologies and funding (Hochstein 2022, p. 8). This accurately underlines that what defines different domains might be a whole host of items. These could be the more epistemically or theoretically focused elements of a "field" in Darden and Maull's sense, including problems, relevant facts, explanatory norms, techniques, methods, concepts, laws, theories, or vocabularies. A more sociological bent could identify them with one of the many accounts of scientific communities available in the history, philosophy, and sociology of science (Shapin 2010), or even with Kuhnian paradigms (Kuhn 1996). But again, the minimal sense of domain is the one that I want to rely upon in what follows—the only necessary feature is that domains are the source for scientific explanations that can exhibit explanatory virtues and, potentially, be integrated with one another.

Let's sum up. For all that a picture of compositional levels appears in nearly every introductory biology textbook, this is extremely difficult, if not outright impossible, to work out in the details. If we have our sights on explanations that integrate these levels, even a number of the philosophical accounts that have been proposed seem to give us precious little way in which to interpret vertical integration. We were left with an extremely general notion of levels drawn from Eric Hochstein (2022), on which differences between levels are about differences in the foregrounding and backgrounding of information used to generate scientific explanations. What, then, is it to *integrate* a level? Again, an assortment of philosophical accounts bequeaths us little clarity. I argued that one particular feature—combining explanations from multiple domains in a manner that exhibits transformative interdisciplinarity—could suffice as a minimal characterization of integration for our purposes.

Applying these insights to vertical and horizontal integration in the biological sciences, then, we saw two things emerge: first, while vertical integration is difficult to defend after the demise of a strict compositional hierarchy, we can nonetheless salvage an account that brings together both vertical and horizontal integration as inter-domain integration, thanks to the productive breadth of Hochstein's presentation of levels.

What of integrative promise, then? As mentioned in the last chapter, I claim that inter-domain integration is *one* of the several explanatory virtues that constitute integrative promise. The possibility of engaging in transformative interdisciplinary work that could unify either disparate scales or disparate clades of the living world is prized by practicing biologists, and explanations that exhibit it tend to be pursued and supported. But integration is not the only feature of promise. As we will see in the following chapters, this central epistemic aim of integration is supplemented in integrative promise by a further set of virtues, to which we now turn.

References

Bod R, Dongen Jv, Hagen SLt, Karstens B, Mojet E (2019) The flow of cognitive goods: a historiographical framework for the study of epistemic transfer. Isis 110(3):483–496. https://doi.org/10.1086/704673

Brigandt I (2010) Beyond reduction and pluralism: toward an epistemology of explanatory integration in biology. Erkenntnis 73(3):295–311. https://doi.org/10.1007/s10670-010-9233-3

Brigandt I (2012) The dynamics of scientific concepts: the relevance of epistemic aims and values. In: Feest U, Steinle F (eds) Scientific concepts and investigative practice. Walter de Gruyter, Berlin, pp 75–103

Brigandt I (2013) Systems biology and the integration of mechanistic explanation and mathematical explanation. Stud Hist Philos Biol Biomed Sci 44(4):477–492. https://doi.org/10.1016/j.shpsc.2013.06.002

Brooks DS (2021) A new look at "levels of organization" in biology. Erkenntnis 86(6):1483–1508. https://doi.org/10.1007/s10670-019-00166-7

Brooks DS, Eronen MI (2018) The significance of levels of organization for scientific research: a heuristic approach. Stud Hist Philos Biol Biomed Sci 68–69:34–41. https://doi.org/10.1016/j.shpsc.2018.04.003

Craver CF (2015) Levels. Open MIND 8(T). https://doi.org/10.15502/9783958570498

Craver CF, Darden L (2013) In search of mechanisms: discoveries across the life sciences. University of Chicago Press, Chicago

Currie A (2019) Simplicity, one-shot hypotheses and paleobiological explanation. Hist Philos Life Sci 41(1). https://doi.org/10.1007/s40656-019-0247-0
Currie A (2021) Comparative thinking in biology. Cambridge University Press, Cambridge
Darden L, Maull N (1977) Interfield theories. Philos Sci 44(1):43–64. https://doi.org/10.1086/288723
DiFrisco J (2017) Time scales and levels of organization. Erkenntnis 82(4):795–818. https://doi.org/10.1007/s10670-016-9844-4
Eronen MI (2015) Levels of organization: a deflationary account. Biol Philos 30(1):39–58. https://doi.org/10.1007/s10539-014-9461-z
Hochstein E (2022) Foregrounding and backgrounding: a new interpretation of "levels" in science. Eur J Philos Sci 12(2):23. https://doi.org/10.1007/s13194-022-00457-x
Huutoniemi K, Klein JT, Bruun H, Hukkinen J (2010) Analyzing interdisciplinarity: typology and indicators. Res Policy 39:79–88. https://doi.org/10.1016/j.respol.2009.09.011
Kästner L (2018) Integrating mechanistic explanations through epistemic perspectives. Stud Hist Philos Sci 68:68–79. https://doi.org/10.1016/j.shpsa.2018.01.011
Kim J (2002) The layered model: metaphysical considerations. Philos Explor 5(1):2–20. https://doi.org/10.1080/10002002018538719
Kitcher P (1981) Explanatory unification. Philos Sci 48(4):507–531
Klein JT (2010) A taxonomy of interdisciplinarity. In: Frodeman R, Klein JT, Mitcham C, Holbrook JB (eds) The Oxford handbook of interdisciplinarity. Oxford University Press, Oxford, pp 15–30
Kuhn TS (1996) The structure of scientific revolutions, 3rd edn. University of Chicago Press, Chicago
Leonelli S (2016) Data-centric biology: a philosophical study. University of Chicago Press, Chicago
List C (2019) Levels: descriptive, explanatory, and ontological. Nous 53(4):852–883. https://doi.org/10.1111/nous.12241
Love A (2008) Explaining evolutionary innovations and novelties: criteria of explanatory adequacy and epistemological prerequisites. Philos Sci 75(5):874–886. https://doi.org/10.1086/594531
Machamer P, Darden L, Craver CF (2000) Thinking about mechanisms. Philos Sci 67(1):1–25. https://doi.org/10.1086/392759
Mayntz R (1992) The influence of natural science theories on contemporary social science. In: Dierkes M, Biervert B (eds) European social science in transition: assessment and outlook. Campus Verlag, Frankfurt, pp 27–79
Neto C (2020) When imprecision is a good thing, or how imprecise concepts facilitate integration in biology. Biol Philos 35(6):58. https://doi.org/10.1007/s10539-020-09774-y
Oppenheim P, Putnam H (1958) Unity of science as a working hypothesis. In: Feigl H, Scriven M, Maxwell G (eds) Concepts, theories, and the mind-body problem. University of Minnesota Press, Minneapolis, pp 3–36
Plutynski A (2013) Cancer and the goals of integration. Stud Hist Philos Biol Biomed Sci 44(4):466–476. https://doi.org/10.1016/j.shpsc.2013.03.019
Potochnik A (2017) Idealization and the aims of science. University of Chicago Press, Chicago
Potochnik A, McGill B (2012) The limitations of hierarchical organization. Philos Sci 79(1):120–140. https://doi.org/10.1086/663237
Schmidt JC (2008) Towards a philosophy of interdisciplinarity: an attempt to provide a classification and clarification. Poiesis Praxis 5:53–69. https://doi.org/10.1007/s10202-007-0037-8
Shapin S (2010) The scientific life: a moral history of a late modern vocation. University of Chicago Press, Chicago
Slater MH (2017) Naturalized metaphysics and the contention over the ontological status of species. In: Slater MH, Yudell Z (eds) Metaphysics and the philosophy of science: new essays, chap 3. Oxford University Press, Oxford, pp 55–80
Thalos M (2013) Without hierarchy: the scale freedom of the universe. Oxford University Press, Oxford

Triviño V (2022) Towards a characterization of metaphysics of biology: metaphysics *for* and metaphysics *in* biology. Synthese 200:428. https://doi.org/10.1007/s11229-022-03897-3
Wimsatt WC (1974) Reductive explanation: a functional account. In: PSA: Proceedings of the biennial meeting of the Philosophy of Science Association, vol. 1974, pp 671–710
Wimsatt WC (1976) Reductionism, levels of organization, and the mind-body problem. In: Globus GG, Maxwell G, Savodnik I (eds) Consciousness and the brain. Springer, Boston, pp 205–267

Chapter 3
Stringency and Opportunism

Darwin's Orchids, Insectivorous Plants, and Earthworms

Abstract In addition to integration, integrative promise also invokes a number of other explanatory virtues. In particular, because integrative knowledge in biology is extremely difficult to come by, promise involves a trade-off between what I will call stringency and opportunism. Stringency refers to the idea that integrative explanations need to be put to severe tests, given the challenges that we face in attaining integration. But at the same time, opportunism describes the tendency of biologists to rely on explanations that use readily available materials, model systems, bodies of theory, and so forth, as a way to make integrative research pragmatically more feasible. In this chapter, I explore this trade-off by looking at the late works of Darwin, who spent much of his career attempting to offer evolutionary explanations with integrative promise using organisms like orchids or earthworms, and struggled mightily with the balance between stringency and opportunism.

Keywords Stringency · Opportunism · Charles Darwin · Orchids · Plant carnivory · Earthworms

As I already hinted at in the preface and the first chapter, part of what makes appeals to integrative promise philosophically interesting is the fact that they seem to merge epistemic with non-epistemic considerations. Having looked at the primary epistemic feature of promise in the last chapter, it's time, in this and the following chapter, to turn to three further important features, whether a mix of epistemic and non-epistemic aspects (in this chapter) or of an entirely non-epistemic character (in Chap. 4).

The nature of the two such features that are our topic here—*stringency* and *opportunism*—can perhaps best be seen when we return to a connection with the extant literature on theoretical virtues that I discussed in Chap. 1. In what way, we might ask, is the pursuit of integrative promise different from the pursuit of *scope*? As we've just seen, promise involves the search for the integration of different domains, which seems like a basic appeal to the virtue of increasing what we called (following Schindler 2018) the *empirical scope* of our theory.

C. H. Pence, *Integrative Promise*, Synthese Library 499,
https://doi.org/10.1007/978-3-031-88471-9_3

However, integrative promise goes beyond scope precisely in its invocation of criteria that combine epistemic and non-epistemic concerns—and particularly in its use of stringency and opportunism, which make the contrast with empirical scope more apparent. I'll elaborate this contrast in general before turning to an especially clear case study of stringency and opportunism from the history of biology: the later works of Darwin. As we'll see, Darwin spent much of his time after publishing the *Origin of Species* in pursuit of integrative promise, and there is no better exemplar of the complex interplay between stringency and opportunism than his efforts to defend his young theory of evolution by natural selection.

3.1 The Pursuit of Empirical Scope

To be sure, biologists do indeed occasionally seem to argue directly for the empirical scope of their theories, or to pursue explanations based on their potential to demonstrate increased empirical scope. These kinds of discussions tend to take two forms.

First are efforts to establish that a given biological theory is genuinely *universal*—that is, that it *must* apply to any living system configured in some particular way. There is a tradition of attempts to understand evolutionary theory as exactly such a "universal" theory, both within the philosophy of biology (see discussion in Pence 2021, pp. 53–62) and in biology itself, especially in the literature on the mathematical modeling of evolution (e,g., Conway Morris 2009; Frank 2017, 2020; Grafen 2014; Rice 2020). If evolutionary theory is really something more like an algorithm, which would apply to any reproducing and differentially varying population of objects (say, any population that satisfied something like the Lewontin conditions, sensu Lewontin 1970), then the precise, maximal limits of the empirical scope of evolutionary theory are known, pending potential future efforts to render those conditions more expansive.

The second kind of argument deployed in direct pursuit of empirical scope seeks to extend the number of biological systems explained by a given theoretical framework, whether that involves the addition of further populations, species, ecosystems, and so on. Again, we occasionally see endeavors that seem to move in this direction. The use of environmental DNA methods—essentially, taking a mass sample from an environment and sequencing all DNA found within it in an attempt to identify species on the basis of DNA barcoding (Shokralla et al. 2012; Yoccoz 2012)—seems to be aimed at improving the empirical scope of ecological theories in its own right, as one often has little or no knowledge of the organisms that will be cataloged in advance. The effort focuses instead, for instance, on offering new ways in which to confirm our extant understandings of trophic interactions or interspecies relationships.

But neither of these features is present, I argue, in explanations that are prized for their integrative promise. Let's take the second sort of defense first. The ends of integrative promise won't be served by indiscriminately searching for further biological systems to which our present theories might apply. As we saw in the last chapter, it will often instead be a matter of integrating domains with extant bodies of biological theorizing, or, as we'll see below, of seeking out empirical cases that have very particular kinds of features. This is quite the opposite, then, of the direct search for empirical scope.[1]

And while one might argue that a theory with universal scope will by definition exhibit some integrative promise—for instance, a "universal" mathematical model for evolution would arguably be able to offer us horizontal integration across a wide variety of biological systems—this result is also often not borne out in the practice of biologists on the hunt for theoretical universality. Generalized mathematical models of evolutionary theory often specifically dismiss the kinds of crucial cases or targeted integrations that we have already discussed. As Richard Levins (1966, p. 422) famously discussed, model-building, especially in population biology, inherently involves a structure of trade-offs. The kinds of universal models we have reviewed have a tendency to, in Levins's words, sacrifice realism to generality and precision. Integrative promise, in the cases we have seen and will continue to elaborate as we move forward, is more likely to place itself in the camp of sacrificing precision to retain generality and realism, appealing not to idealized versions of biological systems but to concrete, real-world cases in the pursuit of inter-domain integration.

Rather than seeking empirical scope directly, then, I argue that explanations advanced for their integrative promise often display two other, quite different features instead: what I will call *stringency* and *opportunism*. In short, by stringency I mean to refer to the fact that integrative-promise explanations must set a very high standard for the cases used for integration; by opportunism I mean the fact that such cases are drawn from those "at hand" for the biologists involved. As we'll see, I mention them together with good reason: they present their own kind of trade-off structure, and the impact that they have on integrative promise is clearest when we consider them together. But to begin, let's see how these two features are expressed in another real-world example of research undertaken for its promise: the later and lesser-known works of Darwin.

3.2 Darwin's Orchids

Just three years after the publication of the *Origin of Species*, Darwin published his next book: *On the Various Contrivances by which British and Foreign Orchids are Fertilised by Insects* (Darwin 1862b). The work is something of a survey,

[1] This search for very particular features of systems is also noted by Angela Potochnik (2017), who writes that casual patterns worth explaining will be "either of intrinsic interest or of interest due to their potential for intervention," and "somewhat simple and somewhat general," yielding "the right degree of specificity to provide the basis for intervention" (Potochnik 2017, p. 42). Elsewhere, she argues directly against the utility of "universal" explanations that encompass any possible causal influence on a system (Potochnik 2017, pp. 151–152).

described in a recent scientific volume as "a rather short study in the comparative floral morphology and biomechanics of, at the time, 28 species (in 15 genera) distributed throughout Britain" (Bernhardt and Edens-Meier 2014, p. 8). As the title indicates, Darwin focuses on their fertilization, with an emphasis on orchids' complex coadaptive relationships with their insect pollinators.

All well and good, but as Jim Endersby notes, this might *a priori* strike us as something of a paradox:

> Given the extraordinary impact of the *Origin*, we must ask why Darwin's follow-up book was a rather dry and technical tome on orchids rather than a best-selling work on human evolution—a topic he had previously avoided and which most of his readers were probably expecting him to finally address. (Endersby 2016, p. 82)

Despite Darwin's unsuccessful efforts to sweep the question of human evolution under the rug in the *Origin* itself—"light will be thrown on the origin of man and his history" (Darwin 1859, p. 488) was the only mention of human beings in the volume—everyone immediately saw that this was the most earth-shattering implication of his work. His analysis of that question, though, would have to wait another nine years, for the publication of *The Descent of Man* (Darwin 1871). There were orchids to describe. And, I contend, explanations of the morphology behind their fertilization seemed to Darwin to be particularly pregnant with integrative promise.

3.2.1 The Birth of an "Orchidaceous Maniac"

Before turning to that story, however, we should start with some context. In the nineteenth century, orchids were everywhere. Endersby's work on the cultural history of orchids (spanning a much wider period) notes that "Darwin's research was helped by the prevailing orchidmania; the high prices paid for rare orchids allowed a whole industry of collectors, shippers, nurserymen, writers, and illustrators to grow up around the flowers" (Endersby 2016, p. 6). He notes, as well, the review of Darwin's orchids volume that appeared in *The Athenaeum*, a popular literary magazine of the day, which puts Darwin's work not in the domain of evolutionary theory (or at least not until the very end, where "the author's well-known peculiar views" get a brief mention; Leifchild 1862, p. 685), but rather in that of "the existing fashionable fancy for cultivating Orchids," the "grotesque attractiveness" of the "simulations of animal forms which Orchids so strangely take," and the behavior of an "orchidaceous maniac" at the local auction-house where rare varieties were bought and sold (Leifchild 1862, p. 683). The circulation of these species—about which more in a moment—was not only a question of scientific networking, but also of trade routes, guarded by the British Navy and extended by the Empire (Endersby 2016, p. 6). The impact of Darwin's *Orchids* was even felt as far away as the world of fiction, where it prompted H. G. Wells to write his short story "The Flowering of the Strange Orchid" ("you know, Darwin studied their fertilisation," our main character tells an orchid-skeptic; Wells 1894).

The occasionally-repeated claim that Darwin's work was some kind of hyper-technical treatise on a subject that almost no one would or could have appreciated at the time is thus clearly off the mark. As Bernhardt and Edens-Meier, two botanists working on orchids, write in introducing a contemporary scientific tribute to the *Orchids* book,

> If a first edition of a scientific work goes on to a second and expanded edition without obvious objection and rejection by the scientific community, it can't be called obscure. If the revised edition continues to be read and cited by laypersons and professional scientists for seminal information, including some protocols still in use today, it can't be called obscure. If the book provides repeated inspiration to authors of imaginative fiction, it can't be considered obscure. (Bernhardt and Edens-Meier 2014, p. 19)

At least in this sense, then, Darwin's work on the orchids takes its place in a relatively long and robust tradition (a tradition equal parts popular and scientific) of orchid-mania.

The fertilization of plants also wasn't a novel source of inspiration for Darwin. As early as 1839, in the years during which he privately worked away on his theory of natural selection, he corresponded on questions of plant fertilization with the botanist William Herbert. Via his colleague and friend John Stevens Henslow, he had written to ask Herbert a handful of botanical questions, and Herbert had offered a few speculations on some results concerning the differential fertility of crosses in *Hippeastrum*. Darwin is intrigued, and writes to Herbert himself for clarification. "I ask," he writes, "because I have been led to believe, that amongst organic being[s] producing seminal offspring there exist[s] no such thing as a true permanent hermaphrodite—i.e. that every individual occasionally, though perhaps *very rarely*, after long intervals is fecundated by another indiv" (Darwin 1839).

Orchids, in particular, became a subject of research around the time that Darwin moved to his home in Downe a few years later, in 1842 (a coincidence to which we we will return). As Michael Boulter writes in his work on the role of Down House in Darwin's theorizing, Darwin and his wife Emma found themselves surrounded by flowers on their rural estate, where "green-winged orchids grew with cowslips in the fields and the young couple brought back pots of the wild plants to transplant in the greenhouse" (Boulter 2008, p. 157). A hillside near their home, which they referred to as "Orchis Bank" (Downe Bank, now protected by the Kent Wildlife Trust), plays host to numerous local species.

The question of fertilization continued to be important as Darwin began to commit his theory of natural selection to paper. In his "Essay" of 1844, one of the earliest drafts of the theory of natural selection, he writes that "such contrivances exist in so very many hermaphrodite flowers by which an occasional cross may take place, that I cannot avoid suspecting (with [the botanist] Mr [Thomas Andrew] Knight) that the reproductive action requires, at *intervals*, the concurrence of distinct individuals" (Darwin 1909, p. 70).[2] Darwin would obliquely refer to "the general

[2] This hypothesis that crossing is not just beneficial but *required* in the long run would go on to be known as "the Knight-Darwin law," and became the subject of a small literature.

law of good being derived from the intercrossing of distinct individuals of the same species" in the *Origin* (Darwin 1859, p. 99). He also was already prepared by that point to claim that "many of our orchidaceous plants absolutely require the visits of moths to remove their pollen-masses, and thus to fertilise them" (Darwin 1859, p. 73).

Darwin was therefore primed for work on orchids, and he began to seriously collect data on their morphology and their relationships with insect pollinators in 1858 (especially on specimens of *Ophrys muscifera* that he had been keeping for some time), describing them in a letter to the *Gardeners' Chronicle* a few years later (Darwin 1860a). In addition, he was on the hunt for fellow naturalists who could help him with the details of a difficult case:

> I should be extremely much obliged to any person living where the Bee or Fly Orchis is tolerably common, if he will have the kindness to make a few simple observations on their manner of fertilisation. […] It is this curious apparent contradiction in the structure of the Bee Orchis—one part … adapted for fertilisation by insect agency—another part … adapted for self-fertilisation without insect agency—which makes me anxious to hear what happens to the pollen-masses of the Bee Orchis in other districts or parts of England. (Darwin 1860a, p. 528)

As it turns out, poor Darwin's query would not be answered satisfactorily until the twentieth century, when it was realized that this orchid was deceiving bees into trying to copulate with it (Endersby 2016, pp. 103–104, 201–202). Let's leave Darwin's milieu aside, then, and turn to the book that resulted.

3.2.2 *The* Orchid *Book*

What exactly do we find in Darwin's treatise on orchids? We don't need to occupy ourselves too much with its details. As James Costa writes, the kinds of descriptions that he offers of the movements of all the parts of orchids when visited by different insects "would be eyeball-glazing to any but the most devoted orchid (or pollination) aficionado" (Costa 2017, p. 243). As Endersby notes, in his review of the volume even the botanist Joseph Dalton Hooker "had to admit 'that Mr. Darwin's is no work for the general reader,' because of the complexity of both orchid flowers and the analysis of their form and function" (Endersby 2016, p. 84).

But we can still consider two particularly emblematic cases that give us a nice encapsulation of the kind of work that Darwin was up to. To start, consider the early-purple orchid, *Orchis mascula* (Fig. 3.1). Well known on English hillsides, Endersby notes that this orchid was quite likely the "long purple" with which Ophelia, in Shakespeare's *Hamlet*, adorned herself before drowning (Endersby 2016, pp. 133–134). This is the first orchid which Darwin describes in detail, and it serves as a perfect example of the kind of coadaptation between insect pollinator and flower that Darwin found so fascinating.

Say that an insect lands on the labellum petal of the flower (labeled *l* and visible in both views *A* and *B* of Fig. 3.2). It will push its head toward the back of the flower,

Fig. 3.1 *Orchis mascula*, the first species of orchid discussed by Darwin and the most prevalent locally, photographed in Belgium. Wikimedia Commons, CC-BY-SA 4.0, by Hans Hillewaert

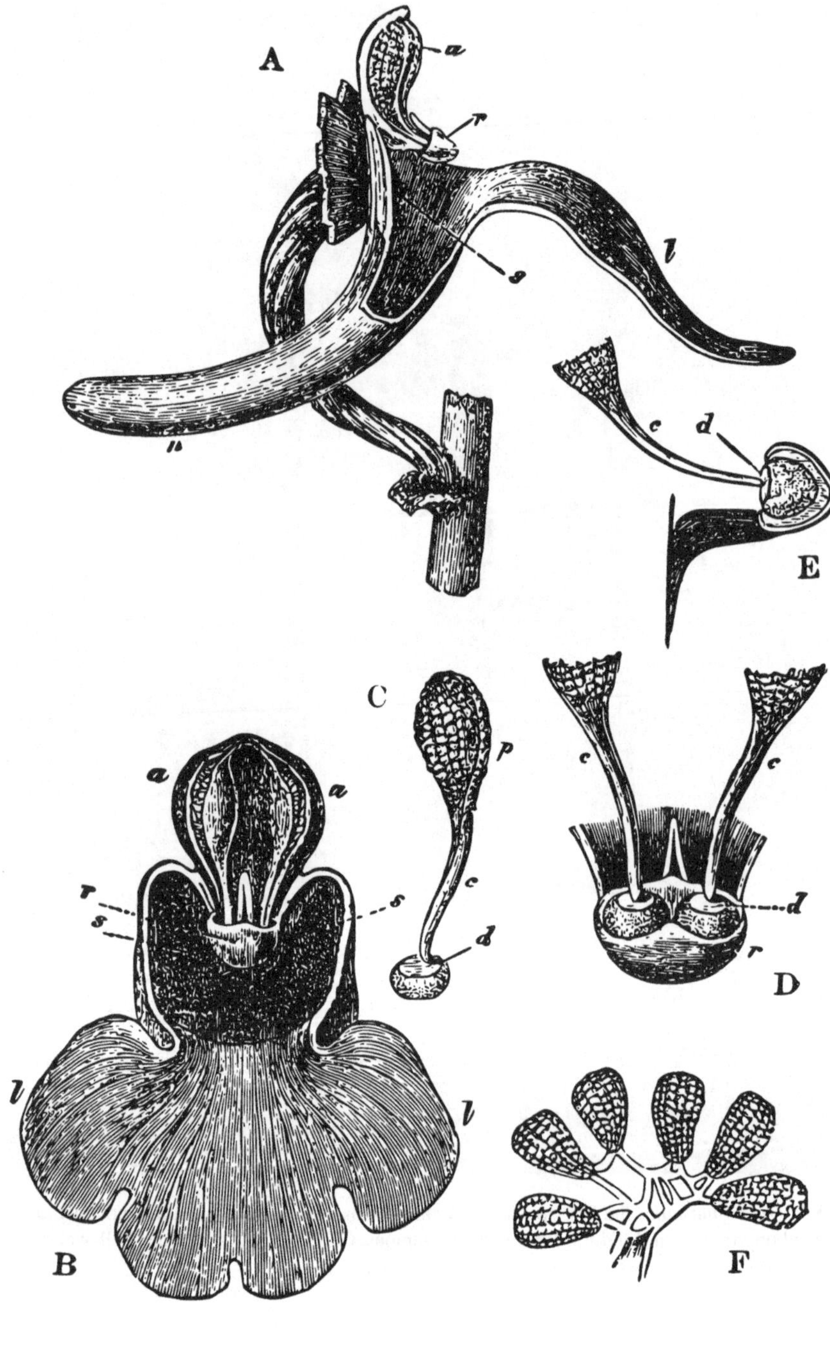

ORCHIS MASCULA.

Fig. 3.2 Darwin's diagram of *Orchis mascula*, from the second edition of *Orchids* (1877, p. 8). Public domain

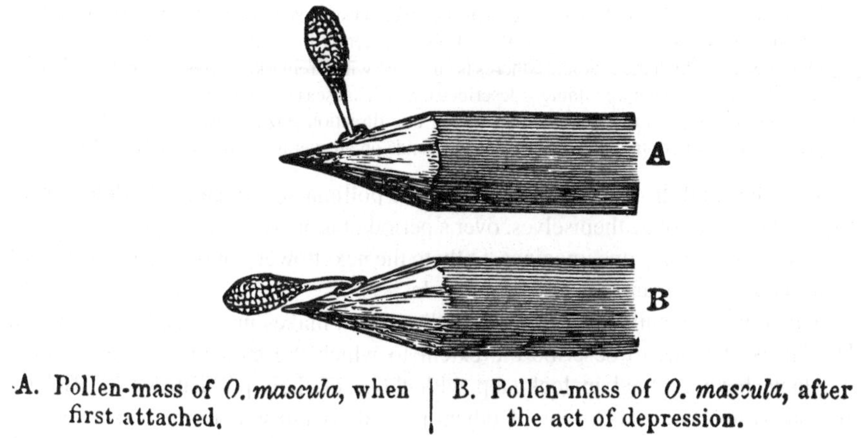

Fig. 3.3 The celebrated "pencil experiment," which Darwin used to argue that orchid pollinaria dispersion was constructed to permit insect cross-pollination (Darwin 1877, p. 12). Public domain

where the stigmas are found (*s* in *A* and *B*) in an effort to reach the nectar in the base of the flower (*n*, at left in *A*). Deftly positioned directly in the way of this process, however, is the rostellum (*r*). But "the slightest touch causes it [the rostellum] to rupture transversely in a sinuous line.... This act makes no difference in the shape of the rostellum, but converts the front part into a lip, which can be depressed easily" (Darwin 1877, p. 10).[3] Illustration *D* in Fig. 3.2 shows what the depression of this lip accomplishes: it reveals two sticky discs (*d*), attached to a stalk (*c*) bearing all of the plant's pollen in two great masses (the pollinia, *p*). "When this is effected," Darwin writes, "one or both of the viscid balls will almost infallibly touch the intruding body. So viscid are these balls that whatever they touch they will firmly stick to" (Darwin 1877, p. 12). Thus the first half of the coadaptive dance: any insect that arrives to grab a sip of nectar will leave with one or two pollinia glued to itself—glue which was protected from setting too early by being stored in the membrane-sealed pouch of the rostellum, opened at the touch of the insect's landing.

But this would only commence the battle. If the pollinia stayed in this initial vertical position (like one or two extra bulbous antennae on our pollinator), then they would impact the wrong part of the plant. Look again at the side-view, illustration *A* in Fig. 3.2: vertical pollinia would impact at point *a*, while the stigma is much lower, at point *s*. It's here that Darwin's second insight is important. When he tested the attachment behavior of the pollinia using a pencil, he found that as the "glue" dried, the pollinia leaned forward to point directly at the location of the stigma (Fig. 3.3). In Darwin's own words,

[3] I will quote from the second edition of the *Orchids* book, which was expanded, corrected, and, as Bernhardt and Edens-Meier (2014) have argued, was the version almost exclusively used by practicing scientists as soon as it was published.

> How then can the flower be fertilised? This is effected by a beautiful contrivance: though the viscid surface remains immovably affixed, the apparently insignificant and minute disc of membrane to which the caudicle adheres is endowed with a remarkable power of contraction (as will hereafter be more minutely described), which causes the pollinium to sweep through an angle of about ninety degrees, always in one direction, viz., towards the apex of the proboscis or pencil, in the course of thirty seconds on an average. (Darwin 1877, p. 13)

Now we have all the pieces of the puzzle: the pollinia are structured both to adhere to insects and to rotate themselves, over a period of around thirty seconds (the length of time it would take, one imagines, to fly to the next flower), into exactly the correct position to fertilize another *Orchis mascula*.

This kind of detailed morphological discussion makes up the bulk of the work. Darwin is at pains to describe the extent to which the structures of fertilization in plants have evolved in lock-step with those of their pollinators. As Richard Bellon writes, "he and dozens of others poured countless hours into the study of complex co-evolutionary relationships between pollinators and flowering plants" (Bellon 2009, p. 398).

Obviously enough, when presented with this sort of "beautiful contrivance," his readers' minds would quickly be drawn to the role of an intelligent designer.[4] Darwin doesn't have the ability to trace out in detail the story of the delicate evolution of coadaptations between insect and flower. In the last edition of the *Origin*, into which he has incorporated a brief discussion of some of his orchid results, he notes that if we follow the evolutionary chain back far enough, "it may be asked how did the stigma of an ordinary flower become viscid, but as we do not know the full history of any one group of beings, it is as useless to ask, as it is hopeless to attempt answering, such questions" (Darwin 1876, p. 196).

What he can do instead, however, is show us cases where evolution's *tinkering* comes to the fore, by contrast with the kinds of explanations that design might offer us.[5] Already, the orchids in general give us reason to think that things are not as simple as they might seem. In most orchids, the labellum petal—the "landing pad" (*l*) that triggered the opening of the compartment containing the sticky discs in our *Orchis mascula*—is a strange thing. Darwin writes that

> In most of the Orchidæ … the position of the labellum is the more remarkable, because it has been purposely acquired, as shown by the ovarium being spirally twisted. In all Orchids the labellum is properly directed upwards, but assumes its usual position on the lower side of the flower by the twisting of the ovarium.… (Darwin 1877, p. 131)

That is, the position of the labellum petal *l* as it is depicted in illustration *A* of Fig. 3.2 is *not* the way that the flower initially opens. As you can see if you look at the twisting that Darwin has indicated in his diagram of the flower's support, the labellum petal begins on the *top* of the flower, where it would do no good at all

[4] Endersby notes that the bee orchis, the example of mimicry that Darwin sought help in understanding above, would be taken up by St. George Jackson Mivart in his largely theologically driven critique of the *Origin* (Mivart 1871).

[5] In detailing the following example, I follow the groundbreaking work of John Beatty (2006).

as a landing site for potential pollinators. After the flower opens, it *twists* through one hundred and eighty degrees until the labellum can play the complex role that is described for it above. It is here that *Malaxis paludosa* enters the scene. This relatively unremarkable "bog orchid" (Fig. 3.4) is impressive for *not* working in this way.

As we can see in the photograph (bottom-center flower, Fig. 3.4) and in Darwin's accompanying diagram (*l* in drawing *A* of Fig. 3.5), in the open flowers of *Malaxis*

Fig. 3.4 *Malaxis paludosa*, (today, *Hammarbya paludosa*), the orchid with a "contrivance" that most clearly distinguishes itself from the work of a designer. Wikimedia Commons, CC-BY-SA 2.0, by Len Worthington

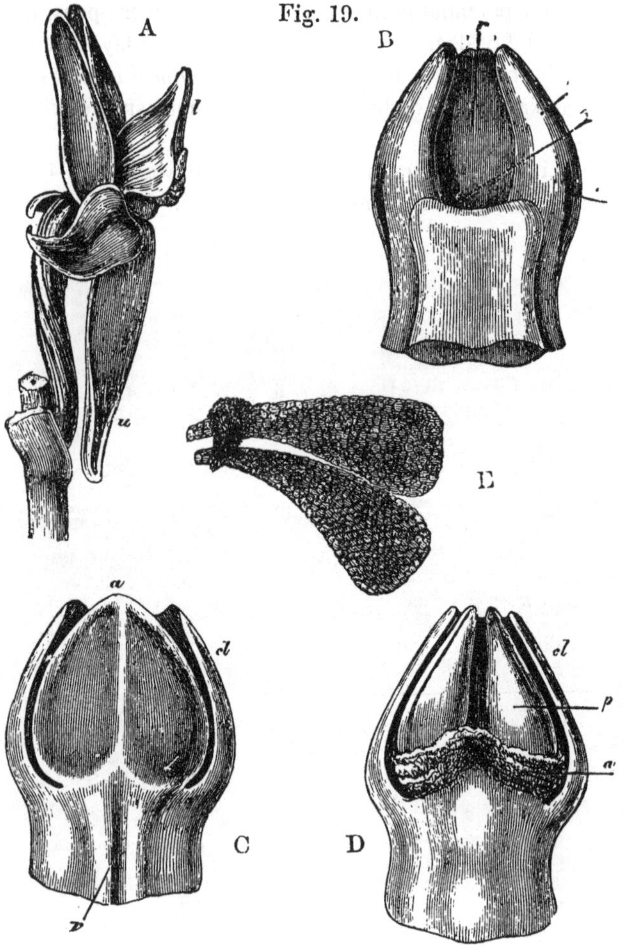

MALAXIS PALUDOSA.

(Partly copied from Bauer, but modified from living specimens.)

a. anther. *v.* spiral vessels.
p. pollen. *r.* rostellum.
cl. clinandrum. *s.* stigma.
l. labellum.
u. the sepal which in most orchids stands on the upper side of the flower.

A. Perfect flower viewed laterally, with the labellum in its natural position, upwards.

B. Column viewed in front, showing the rostellum, the pocket-like stigma, and the anterior lateral portions of the clinandrum.

C. Back view of the column in a flower-bud, showing the anther with the included pear-shaped pollinia dimly seen, and the posterior edges of the clinandrum.

D. Back view of an expanded flower, with the anther now contracted and shrivelled, exposing the pollinia.

E. The two pollinia attached to a little transverse mass of viscid matter, hardened by spirits of wine.

Fig. 3.5 Darwin's diagram of *Malaxis paludosa*, from the second edition of *Orchids* (1877, p. 130). Public domain

paludosa, the labellum petal is actually once again on the top. Yet another complex contrivance—this time involving coating a visiting insect in a less firmly setting glue and withdrawing the freely-dangling pollinia when it exits (Darwin 1877, p. 134)—ensures that, nonetheless, insect visitors will still serve as effective cross-pollinators for this orchid as well.

But what makes this orchid useful for Darwin's purposes is that it does not behave in the way that one would expect if it were intelligently designed. A designer, Darwin assumes, would be expected to follow something like Ockham's razor—since the flowers first emerge with their labellums on the upper side, then twist, why not simply leave the labellum on top, should that be the adaptively preferable solution in the given environment? But that's not what *M. paludosa* does. Rather,

> in Malaxis the twisting has been carried so far that the flower occupies the position which it would have held if the ovarium had not been at all twisted, and which the ripe ovarium afterwards assumes, by a process of gradual untwisting. (Darwin 1877, p. 131)

That is, the labellum petal of *Malaxis paludosa* starts out on the top, twists through one hundred and eighty degrees to the bottom (like every other orchid), then *continues twisting* until it has spun a full three hundred and sixty degrees and arrived back where it started. Then, once the flower has been fertilized and seed has started to set, it *untwists itself*, back through the entire full rotation, to end the way that it started when it opened.

Darwin doesn't leave the explanation for this in terms of natural selection as an exercise for the reader. As he introduces the discussion in the book's conclusion,

> In my examination of Orchids, hardly any fact has struck me so much as the endless diversities of structure,—the prodigality of resources,—for gaining the very same end, namely, the fertilisation of one flower by pollen from another plant. This fact is to a large extent intelligible on the principle of natural selection. As all the parts of a flower are co-ordinated, if slight variations in any one part were preserved from being beneficial to the plant, then the other parts would generally have to be modified in some corresponding manner. But these latter parts might not vary at all, or they might not vary in a fitting manner.... (Darwin 1877, p. 284)

The comparison to design isn't made explicit, but the positive case for selection certainly is. Natural selection makes it entirely expected that the "problem" of insect fertilization would be solved in a vast number of ways by different species of plants. Partly this is due to the rarity of the initial appearance of the extraordinary, favorable variation that would give the plant an advantage—the first bit of adhesive material at the bottom of the pollinia, say. But it's also a result of the fact that variations need to be integrated into the organism as a whole (one is reminded of the architectural constraints that would much later be emphasized by Gould and Lewontin 1979). Variation in one part needs to be accommodated, and if the accommodating variation in other parts isn't available, a variation that might in another context be an adaptation could be harmful. And *Malaxis paludosa* makes a perfect example. Its top-position labellum

> might be simply effected by the continued selection of varieties which had their ovaria less and less twisted; but if the plant only afforded varieties with the ovarium more twisted,

> the same end could be attained by the selection of such variations, until the flower was turned completely round on its axis. This seems to have actually occurred with *Malaxis paludosa*.... (Darwin 1877, pp. 284–285)

As he concludes the argument after providing a few more examples, "the contrivances and beautiful adaptations slowly acquired through [natural selection] transcend in an incomparable manner the contrivances and adaptations which the most fertile imagination of man could invent" (Darwin 1877, pp. 285–286).

We'll return to the significance of both of these lines of argumentation for the rest of Darwin's theorizing in a moment, but before leaving the volume, it's worth noting that these are not the only novelties found in its pages. Bernhardt and Edens-Meier chronicle a whole collection, including Darwin's having been the first to demonstrate the possibility of certain kinds of cross-pollination (i.e., in orchids which had previous been thought to be entirely self-fertilizing); a famous case in which Darwin showed that "plants originally placed in separate genera, due to radically different floral architecture, were actually males and females of the same species"; predictions that certain floral architectures required the existence of other (as-yet undiscovered) pollinator species (a prediction later expanded upon and now associated rather with Alfred Russel Wallace; Beccaloni 2017); and a speculative homology between the (highly unusual) flowers of orchids and those of other flowering plants (Bernhardt and Edens-Meier 2014, pp. 10–13).

3.2.3 *Why Orchids, Why Now?*

What, then, is the resolution to the riddle that Endersby posed earlier—why would Darwin have wanted to write this book at this time? What roles did these arguments surrounding orchids play in his broader defense of natural selection and common descent?

There is, of course, unlikely to be a single answer to such a question. Darwin was fighting multiple battles on multiple fronts, and the kind of analysis taking place in the orchids book could be read as contributing to many of them. Before introducing the sense in which I think this work provides an excellent example of a set of explanations demonstrating the integrative promise of natural selection, I want to discuss some of these other possibilities.

First is a purely epistemic worry. Darwin had faced a number of critiques concerning what we might call the justificational structure of natural selection. Despite the fact that he believed that he had built a theory that satisfied all the canons of good nineteenth-century science (Pence 2018), the *Origin* was admittedly short on observational or experimental detail—Darwin himself always called it an abstract of a larger work that would never, in the end, be published. Crucial here is an argument over the nineteenth-century definition of "induction." If "good science," in the tradition stretching from Bacon to Newton, is supposed to be inductively grounded inferences to natural laws on the basis of observation, then how could

Darwin's theory—which had never recorded a direct observation of the creation of a new species, and was elaborated without knowledge of the processes that led to the appearance and distribution of variation in the wild—be appropriately justified? As the esteemed geologist Adam Sedgwick, one of Darwin's erstwhile friends and mentors, would write in an anonymous review of the *Origin,*

> But I must in the first place observe that Darwin's theory is not *inductive,*—not based on a series of acknowledged facts pointing to a *general conclusion,*—not a proposition evolved out of the facts, logically, and of course including them. To use an old figure, I look on the theory as a vast pyramid resting on its apex, and that apex a mathematical point. (Sedgwick 1860, p. 285)

Whatever relationship to the facts a good scientific theory was supposed to have (though this itself is not particularly clear in Sedgwick's rather elliptical statement), Darwin's theory, it was claimed, didn't have it.

Such criticism was dubious, to be sure. It's not clear to what extent Newtonian mechanics itself actually had worked in this way. That lack of clarity was made all the worse by the fact that the wave theory of light was in its ascendancy at the time—with its crucial role for the ether, apparently just the sort of hypothesized substance that *wasn't* founded on straightforward inductive inferences from observational data. Darwin himself could do nothing but vent to his friend Henslow in a letter:

> I wish, if you ever talk on subject to him [Sedgwick], you would ask him whether it was not allowable (& a great step) to invent the undulatory theory of Light—ie hypothetical unduluations in a hypothetical substance the ether. And if this be so, why may I not invent hypothesis of natural selection ... & try whether this hypothesis of natural selection does not explain (as I think it does) a large number of facts in geographical distribution—geological succession—classification—morphology, embryology &c. &c. (Darwin 1860b)

We could certainly read the work on orchids as attempting to provide precisely the kind of data, then, that Sedgwick and others had argued was lacking. As Bellon puts it, as he began to work on orchids, Darwin "would soon realize that his botany provided a weapon to counter-attack the destructive, widespread criticism that, in writing the *Origin*, he placed himself in opposition to the proper inductive practice of science" (Bellon 2009, p. 378).

Another use to which the book could be put—and one which Darwin himself discussed in a series of letters with the American biologist Asa Gray—was as a way to bolster the arguments in favor of natural selection that he had provided elsewhere. It certainly seemed to have that effect on at least some readers. The botanist George Bentham (nephew of philosopher Jeremy), had started a skeptic of natural selection, but in mid-1862 had begun to come around when presented with Darwin's careful observational methodology. Gray wrote in July 1862 that he was "amused to see how your beautiful *flank*-movement with the Orchid-book has nearly overcome his opposition to the Origin" (Gray 1862a). Darwin was thrilled. "Of all the carpenters for knocking the right nail on the head," he replied a few weeks later, "you are the very best: no one else has perceived that my chief interest in my orchid book, has been that it was a 'flank movement' on the enemy" (Darwin 1862a).[6]

What exactly was this "flank movement?" While, as already noted, Darwin doesn't offer us (nor could he have offered us) any explicit selective stories for how the "contrivances" in these orchids came about, his presentation is such that one implicitly sees that they could have developed by slow and incremental changes, as pollinator and flower became increasingly adapted to one another (see Darwin 1876, pp. 195–196). More importantly, as Darwin spells out across the later editions of the *Origin*, these orchids offer supporting arguments for various claims that Darwin had made but not fully substantiated in the first edition of the book. The extent to which cross-pollination is important buttresses his argument that hybridization is a significant driver of increased variation, and hence increased action of natural selection (Darwin 1876, pp. 238–239). As we have just seen, the diversity of methods by which it is achieved (some of which seem entirely haphazard) offers us an argument for the "tinkering" of natural selection, for Darwin's vision of the constrained role of chance, and against design (Darwin 1876, pp. 154–155; Beatty 2006; Pence 2022, pp. 13–16). They offer us evidence that apparently unimportant or trivial details of morphology might serve important selective purposes (Darwin 1876, p. 172).[7] In short, the orchids serve to bolster a whole host of points on which critics of natural selection might be and had been skeptical.

3.3 Promising Orchids

What about integrative promise, then? I think that part of the reason Darwin pursued his orchid research was precisely because explanations for the characters of these orchids in terms of natural selection seemed to him to present integrative promise. First, though this isn't my focus here (see Chap. 2), they offered Darwin inter-domain integration, this time in the sense of horizontal integration. Evolutionary theory was supposed to give unified, selective explanations for features across the tree of life, and so its capacity for this sort of integration was crucial. As I argued in the last chapter, this search for horizontal integration offers us a different pattern of pursuit of investigations than we would see if the target was only unification or scope. (And as I'll detail below, this horizontal integration also served a rhetorical

[6] This became the launching point for the celebrated correspondence between Darwin and Gray on the question of design, with Gray arguing for a kind of theistic evolution, concluding that the case of the orchids "opens up a knotty sort of question about *accident* or *design*, which one does not care to meddle with much until one can feel his way further than I can" (Gray 1862b; Lennox 2010).

[7] This is the argument cited as the most important by Darwin's son Francis in his biography of his father: "They supplied an argument against those critics who have so freely dogmatised as to the uselessness of particular structures, and as to the consequent impossibility of their having been developed by means of natural selection" (Darwin 1887, p. 3:254).

function, convincing a number of naturalists, zoologists, botanists, and others that evolutionary theory was a viable and workable source of explanations within their domain.) This was a characteristic of Darwin's entire post-*Origin* career, as he argued that features of orchids, insectivorous plants, earthworms, and climbing plants all were susceptible to the same kinds of selective explanations.

3.3.1 Opportunism

But more significantly for the purposes of this chapter, Darwin's work on the orchids exemplifies the relationship between stringency and opportunism that, I argue, typifies appeals to integrative promise. First, Darwin had clearly been opportunistic in his use of orchids as a model. As I noted above, he had turned to them as a model system after moving to his new home in Kent, and noticing that a nearby hillside was home to many of the same species that he would later use as examples, especially in the first edition of *Orchids*. Some of his early experiments with orchids were on those that he found directly in his walking path around his estate. Costa writes that "ever the opportunistic experimentalist, he decided to perform an insect-exclusion experiment, covering some of them [perhaps *O. mascula*?] and leaving others exposed to visiting insects" (Costa 2017, p. 232). As attested by his son Francis in the *Life and Letters of Darwin*, "he was probably attracted to the study of Orchids by the fact that several kinds are common near Down" (Darwin 1887, p. 3:262), and Bernhardt and Edens-Meier confirm that this remains "the most plausible explanation for Darwin's orchidophilia" (2014, p. 4).

Orchids thus began as a fixture of Darwin's everyday environment, one which he believed could serve to help support horizontal inter-domain integration for evolutionary theory—that is, these explanations were *opportunistic*. As I've already noted, this is in some sense an unsurprising factor in biological experimentation and theorizing: to paraphrase Donald Rumsfeld, we go to war with the model organisms and experimental systems we have, not the model organisms and experimental systems we want. But two broader aspects of the invocation of opportunism are important to note here, which is why I've underlined it as one of the components of promise.

First, I take opportunism to offer an effective response to worries about this kind of experimentation that have been raised by Ulrich Krohs. He argues, with respect to contemporary "convenience experimentation"—the increasingly common practice of performing standardized experiments as available opportunistically in kits from major scientific suppliers—that we ought to be worried that such experiments don't allow us to question the more fundamental hypotheses behind the experiments present in those kits. As he puts it, such a convenience experiment "cannot produce any other data than what satisfies the precondition" as "the data are simply not of a sort that would allow for any deviating outcome" (Krohs 2012, p. 55). One might thus question whether opportunism is a virtue at all.

But here, opportunism is just one factor in a much broader, holistic evaluation of a putatively promising explanation. To be sure, not every explanation in the life sciences exemplifies integrative promise. But for those that do, opportunism is just one feature that makes a promising explanation more testable (and, hence, more promising). Counterbalancing features like stringency, among others, should allow us to avoid the kind of problematic outcomes that Krohs warns us against.

Second, because integrative promise is so difficult to find, we have to recognize the role of external, non-epistemic drivers of the ways in which it is sought. Opportunism is one of the primary such drivers in almost all life-science research. In that sense, it's important to note that opportunism isn't only about what's at hand in one's environment, or what one has the resources to work on at the moment. It also draws our attention to the social processes that underlie and permit the expansion of scientific research. In Darwin's case, this involved, perhaps most strikingly, the extensive network of fellow naturalists upon whom he could call to obtain further specimens, especially from outside Britain. The second edition of the *Orchids* book radically expands the coverage of tropical and Australasian orchids, in large part because of this network. Even for the first edition, "Darwin attempted to enlist other naturalists to make observations on orchid flowers native to Britain" (Bernhardt and Edens-Meier 2014, p. 6), and one of the orchids Darwin studied (and of which he received two specimens in 1862) came as a gift from the wealthy horticulturalist James Bateman, at a time that the plant would have fetched £20 a piece (somewhere between £2000 and £19,000 in today's currency, depending on how one makes the conversion).

Opportunism thus draws our attention to the social embedding of scientific practice. Successful integrative explanations require that one is in the right place at the right time, furnished with the right kinds of materials, experience, and training.[8] Darwin certainly was: he had the social and material capital that he needed to carry through his research into orchids, and as it turned out they were indeed able to serve as an important element of a broadening of the evidential base for evolution: orchids really did have promise after all.

3.3.2 Stringency

But this mention of the success of Darwin's work on orchids in actually demonstrating integrative promise is our first hint that opportunism is, so to speak, only one side of the coin. While we want to take into account the fact that the pursuit of integrative promise seems shaped by what's opportunistically available in a particular researcher's experimental and social context, we also have to take account of the fact that sometimes that opportunistic use of local resources simply

[8] Similar points are underlined in the discussion of interfield theories by Darden and Maull (1977); see Chap. 2.

won't work. We'll see some cases of this explicitly in the next section, but in the meanwhile, we can ask why that might happen in the first place. The answer is clear enough: inter-domain integration is an extremely challenging thing to construct. Of all the potential cases that could help us demonstrate it, most of them will likely fail—they won't really offer the kind of support we need to justify the integrative promise of the explanations at issue. In short, we have to be extremely strict in selecting which cases will actually properly support promise—what I'll call *stringency* plays a very important role in integrative promise.

Let's see how this plays out for the orchids. I've argued above that selective explanations of their features were, Darwin thought, high in integrative promise. They could show that traits across the tree of life, even in peculiar groups like orchids, and even when the "contrivances" at issue appeared more likely to arise by design, would make sense on the assumption of natural selection. The case study was chosen opportunistically: the best explanation for Darwin's choice of orchids was their proximity to Down House.

But several characteristics of selective explanations for the traits of orchids make those explanations particularly compelling, and here's where the notion of stringency becomes important. (I don't intend to produce an exhaustive characterization of stringency, and it will likely manifest different features in different explanatory contexts; the discussion that follows is, I hope, detailed enough to be illustrative.) Orchids were a taxon that might initially seem to *resist* explanation by natural selection—in that sense, stringency is sometimes a question of finding especially difficult examples or putting one's explanations to severe tests. Coadaptation remains a complex evolutionary phenomenon even by today's standards (after all, one has to offer a story of the interaction of multiple adaptations in multiple species, not just one), and the identities and behaviors of the pollinator species weren't entirely known when Darwin prepared his analyses. In describing *Orchis mascula*, for instance, Darwin tells us that as he wrote the first edition, "I had not seen any insects visiting the flowers of the present species; but a friend watched some plants, and saw them visited by several humble-bees, apparently *Bombus muscorum*" (Darwin 1877, p. 15). Before he could have known whether natural selection might explain traits in these orchids, Darwin had indications that those explanations would well exemplify stringency, thus potentially offering us lots of integrative promise.

Second, many of the extremely refined adaptations which Darwin cites in the *Origin* are found in animals: his section on "organs of extreme perfection" makes reference to eyes and lungs, and talks about the division of labor as a potential mechanism for their appearance over evolutionary time (Darwin 1859, pp. 186–192).[9] Showing not only that, on the one hand, complex and highly refined adaptations are found in all taxa, and, on the other, that these adaptations might

[9] He also makes reference to an adaptation in barnacles; drawing out the connections between the barnacles work and other invocations of integrative promise is an interesting possibility that I lack the space to pursue here.

be found in parts (like the precise configuration of the petals of orchid flowers) that otherwise seemed to have no practical use (the orchids being apparently perfectly able to self-fertilize), could thus, once again, reinforce the kind of horizontal integration for which Darwin spent so many years arguing.

Last but not least, we should consider the fact that, just like opportunism, stringency is also affected by social and practical context. Stringency is not just about which systems will be epistemically best-positioned to support a case for integration. That case also has to be made rhetorically and publicly, and therefore questions of a researcher's broader community and the reception of the explanations at issue will enter into those explanations' stringency. Botany had always been important to Darwin, not only because he had trained in it as a young naturalist (Sloan 2009, pp. 26–27), but also because some of his influential allies in the early defense of evolution were botanists (such as his friend and mentor Henslow) and botany itself had become an active place for evolutionary debate (the botanist George Bentham's change of mind on evolution, mentioned above, was delivered as an address to the Linnean Society). Convincing botanists was itself an important part of the reception of Darwin's theory, and thus horizontal integration toward the plant kingdom would have been particularly tactically valuable at the time—such cases would have gone farther than average in a defense of selection's integrative promise.

Putting it all together, then, I have emphasized stringency as an important factor in integrative promise because it draws our attention to the care with which the relatively rare instances of stringent corroboration for inter-domain integration are chosen by practicing scientists. Whether this stringency is directly in support of that integration—as when we look for particularly severe tests—or rather in service of the acceptance of those explanations—as when Darwin tries to build good-will among the botanists—we often see very explicit efforts to justify these choices and to carefully craft defenses of integrative promise.

This discussion does, however, highlight an unusual and extremely important feature of the relationship between opportunism and stringency. One might have already observed that, perhaps paradoxically, opportunism and stringency seem to be pushing in directly opposite directions: if we're being particularly cautious with our selection of cases for integration, why would we simply use whatever we have at hand? But two arguments compel me to claim that these are indeed both features that it is vital to emphasize in an understanding of integrative promise. First, as previously noted, they have in common that they are not just the pursuit of increased empirical scope. That is, rather than moving us indiscriminately toward further cases that could fall under the umbrella of natural selection, stringency and opportunism are both about the kinds of factors that matter when biologists engage in cautious selection of case studies.

Second, and more fundamentally, there's a sense in which they nonetheless frequently work together in cases of explanations with integrative promise. Opportunism and stringency seem to be involved in a kind of trade-off structure, a trade-off that will help us understand why claims of integrative promise sometimes fail to pan out. Picking systems that are highly opportunistic will mean that we will regularly fail to find a system that puts integration to a stringent test. Hunting for the stringency-optimal system will occasionally result in the necessity to do research that is opportunistically impossible. Successful appeals to integrative promise, then, require that we happen to find a system that can give us both. It's for this reason that I think it's important to explicitly include stringency in the account—since opportunism and stringency have this reciprocal relationship, including one requires that we include the other.

Consider the concrete case of the orchids. As it turned out, the orchids which Darwin (opportunistically) found at hand on Orchis Bank in Kent were (stringently) the perfect kind of system for demonstrating the evolution of precise coadaptations—better, even, than Darwin could have appreciated in his day, as we now know (Edens-Meier and Bernhardt 2014). We can generalize the point. Explanations that exhibit true integrative promise are hard-won; attempts to find them are unlikely to succeed. They are epistemically risky, as we saw in the last chapter. But they must nonetheless start from the basis of extant scientific theory and practice. Thus, explanations that *really do* exhibit integrative promise are those for which this gamble pays off—those explanations for which the opportunistic effort to secure that integrative promise results in the choice of a particular case that actually does prove useful for the integration of multiple domains. This interaction between inter-domain integration, stringency, and opportunism is thus essential to the understanding of what makes successful promising explanations possible—and of why they are so rare in the first place.

3.4 Insect-Eating Plants and Earthworms

Orchids weren't the only place where Darwin fought to demonstrate the integrative promise of the theory of evolution by natural selection, and I want to end the chapter by mentioning two more—not least because they show us some of the potential limits of integrative promise. The analysis here of both opportunism and stringency indicates that they play a crucial role not only in successful demonstrations of integrative promise, but also in cases where those demonstrations are less than conclusive.

In 1875, Darwin published a tome on what he called the "insectivorous plants"—creatures like the Venus flytrap (native to wetlands in the Eastern United States, and still something of a novelty in England at the time) and the sundew, which seem to take nutrients by trapping and eating passing insects. As James Costa details, the work that led to the publication of this book was perhaps even more opportunistic than Darwin's usual method. Darwin found himself on vacation in a place where sundews might grow (Fig. 3.6), and guessed that the search for nitrogen-bearing compounds would be the reason for their carnivorous habits, their local environment being particularly nitrogen-deprived. As Costa relates the story,

Fig. 3.6 *Drosera rotundifolia*, a carnivorous sundew, which Darwin studied using everything but the kitchen sink. Wikimedia Commons, CC-BY-SA 2.0, by Scott Darbey

> If the substances selected for his nonnitrogenous trials seem a bit haphazard—gum arabic, sugar, starch, diluted alcohol, olive oil, even an "infusion and decoction" of tea—that's because they were: Darwin used whatever was at hand, pilfering from the kitchen and washroom medicine cabinet. (Costa 2017, pp. 270–271)

He was right, in the end: the plants devoured whatever seemed to have the slightest trace of nitrogen.

For Darwin, what makes this case interesting is the analogy that it lets him draw between the powers of digestion in animals and those that seem to be exhibited by these plants:

> There can hardly be a doubt that all the plants belonging to these six genera have the power of dissolving animal matter by the aid of their secretion, which contains an acid, together with a ferment almost identical in nature with pepsin: and that they may afterwards absorb the matter thus digested. (Darwin 1875, p. 356)

Darwin is searching here for yet another kind of integration: in this case, showing that the lines between the animal and plant kingdoms are blurrier than one might have thought—and hence that evolutionary explanations connecting both to common ancestors are also more plausible than one might have assumed. He makes the case even more explicitly early on in the volume:

> A plant of Drosera, with the edges of its leaves curled inwards, so as to form a temporary stomach, with the glands of the closely inflected tentacles pouring forth their acid secretion, which dissolves animal matter, afterwards to be absorbed, may be said to feed like an animal. (Darwin 1875, p. 18)

In this sense, Darwin is hoping that the insectivorous plants will serve as a way to demonstrate evolutionary theory's integrative promise, displaying commonalities between the kinds of explanations on offer in the plant and animal kingdoms.

Unlike in the case of his orchids, however, it turns out that the opportunistic use of *Drosera* and its relatives doesn't produce a particularly enlightening source of integrative promise for evolutionary explanations. The set of plants described by Darwin that secrete digestive juices don't seem to have a common ancestor, so it appears as though this ability has to be a case of convergent evolution, rather than a single transition from a non-digesting ancestor (Darwin 1875, pp. 361–362). That means that he'd need a disunified evolutionary story for each of the relevant lineages—one which he offers, but about which he's somewhat hesitant, perhaps, even, because it is no longer a single, integrative explanation. The case is worse for the power of movement, which is another, maybe more striking, instance of analogy between the plant and animal kingdoms. "Little light," he writes, "can be thrown on the gradual acquirement of the third remarkable character possessed by the more highly developed genera of the Droseraceæ, namely the power of movement when excited" (Darwin 1875, p. 364). While it seemed intuitively to Darwin that plants with the ability to move and digest had to represent some kind of link between the plant and animal kingdoms, he just didn't have the resources to carry the explanations through. The promissory note of integrative promise here simply couldn't be cashed.[10]

Much the same was true for Darwin's last book, now commonly called *Earthworms* but more accurately known as *The Formation of Vegetable Mould, through the Action of Worms, with Observations on their Habits* (Darwin 1881). R. B. Freeman's list of short bibliographical summaries of each of Darwin's works calls this book "outside the main stream of Darwin's work," insofar as it "reverts to his earlier geological interests" (Freeman 1977, p. 164). But I think James Costa makes the better argument, aptly drawing our attention to the fact that while Darwin assuredly talks about worms' power to move soil, and hence to be agents of slow geological change, he explicitly tells us that "I became interested in them, and wished to learn how far they acted consciously, and how much mental power they displayed" (Darwin 1881, p. 3). As he details, Darwin engages in an array of household experiments on worms, chronicling their responses to light, sound, heat, vibration, and so forth (Costa 2017, ch. 9), all in an effort to see whether or not we might find, once again, the erosion of barriers between taxonomic groups—this time not plants and animals, but rather between the thinking and the non-thinking, the cognitive and the instinctive.

Darwin makes the case across a number of examples. The varied responses that an earthworm has to light seems to indicate that—just as we might be able to discover with respect to the behavior of horses, in his example—the animal has certain kinds of mental states that condition its other reactions. "The comparison here implied," he argues, "between the actions of one of the higher animals and of one so low in the scale as an earth-worm, may appear far-fetched; for we thus attribute to the worm attention and some mental power, nevertheless I can see no reason to doubt the justice of the comparison" (Darwin 1881, pp. 24–25). In particular, worms' aptitude at lining the walls of their burrows and plugging up the open ends with specially selected material from their surrounding environment impressed Darwin. Of course, he writes in the conclusion, we should not expect an animal with apparently only rudimentary powers of vision, and hence a very limited experience of the world around it, to have highly refined mental capacities. "But it is far more surprising that they should apparently exhibit some degree of intelligence instead of a mere blind instinctive impulse, in their manner of plugging up the mouths of their burrows" (Darwin 1881, p. 312).

Again, I think it's profitable to read this case in much the way we did that of the insectivorous plants. At least one of Darwin's goals in writing the *Earthworms* book is to show that evolutionary theory can give us an integrated account of the development of mental properties in the natural world. While he does indeed show that there's more mental capacity in earthworms than might meet the eye, it doesn't

[10] As we now know, with good reason: the evolutionary basis of plant carnivory has only been very recently understood, and has more to do with the reuse of common genes for flowering plant defense rather than any homologies with the animal kingdom (Hedrich and Fukushima 2021).

seem as though they're a useful case for helping us understand the nature of the evolution of mental capacities. Opportunism, certainly, but not very useful in terms of stringency.

3.5 A Difficult Trade-Off

As we've seen in this chapter, opportunism and stringency form an important pair of hybrid epistemic and non-epistemic features that make up integrative promise. Supporting the integrative promise of a set of explanations—here, the ability of natural selection to explain the contrivances of orchids, or to show us the gradations of the evolution of digestive capacities in plants and animals or the development of early forms of cognition—is a challenging endeavor. The ways in which a scientist like Darwin builds that support therefore hinge crucially on the twin pragmatic features of opportunism and stringency.

One might object that opportunism is a trivial characteristic that isn't worth the attention that I'm giving it here. Whether in observational cases like those in this chapter, or in experimental contexts, scientists must always begin developing explanations where they are—in their particular epistemic, technical, experimental, and social contexts. In that sense, any attempt to argue for the integrative promise of an explanation will necessarily be opportunistic. If this were all that was meant by opportunism, then it would be plausible to object that opportunism is an uninteresting, nearly necessary feature of scientific work.

But I think such an objection would be misguided—opportunism isn't a benign or uninteresting point. As we've seen in Darwin's case, highlighting opportunism as an element of integrative promise helps us see the ways in which those contexts (like Darwin's wealthy connections at the peak of the British Empire) have material effects on the nature of the proffered explanations. He wouldn't have been able to demonstrate the integrative promise of a natural-selection explanation for the features of his orchids, for instance, in the absence of the kind of social and economic capital that he had at his disposal. For this reason—among others, as we will see in the following chapter—integrative promise is intimately tied up with the social and cultural embedding of science.

But this opportunism also places scientists in the bind of a painful trade-off. For while opportunism is inherent, in some fashion, to scientific practice, the explanations thus developed might not actually have the features required in order to demonstrate integrative promise. Integrative promise involves, as I've dubbed it here, a strict kind of stringency—an argument for promise will probably only be successful with very careful choices of cases deployed for integration. As we saw above, that trade-off came back to bite Darwin on occasion. Orchids were extremely useful. They did indeed exhibit complex coadaptations between insect pollinator and flower structure. The orchids showed a kind of evolutionary tinkering that helped numerous readers understand how something that seemed like a very precise adaptation might be generated and maintained in the absence of an intelligent designer. References to the orchid work were added throughout later editions of the *Origin*, and, as we saw, it was taken to be an important part of his case for natural selection by numerous contemporary readers. That, I've argued, starts to look like a real case for integrative promise.[11]

But we can see as well in the trade-off between opportunism and stringency times when hints of integrative promise fail to work out in the end. Darwin had hoped to offer us promising stories of the evolution of insectivorous plants—blurring the lines between the plant and animal kingdoms—and of cognition in earthworms—blurring the lines between instinct and cognition—and neither of these explanations actually seems to bear the integrative promise that Darwin thought they did. Opportunism offered what he presumed would be apt systems for study, but they failed to meet the rigor of stringency.

Once more, it bears repeating that I do not mean to say that criteria like opportunism and stringency haven't been considered already in the history and philosophy of science. As we've just seen, Darwin's opportunism is, to some extent, the topic of the entire book-length study on his backyard experimentation by James Costa (2017)—Darwin was obsessed with his local environment, and he found under every rock and around every corner some biological feature of interest, managing to turn most of them into arguments in favor of evolution by natural selection. Stringency (and its closely related cousin rigor, which I have eschewed in order to avoid its mathematical connotation) is also a well-understood feature of the scientific process. To take just one particularly salient example, the process of choice of model organisms has received extensive discussion in the philosophical literature (see Chap. 7, as well as, e.g., Green et al. 2018; Dietrich et al. 2020; Ankeny and Leonelli 2020)—a process that obviously relates to my sense of stringency here.

What makes these features interesting in this context is the crucial role that they play in appeals to the integrative promise of some explanation or set of explanations. In addition to the important epistemic features of those explanations (here, the actual fact of their integrating different domains of biological knowledge, as discussed in Chap. 2), to fully understand and evaluate appeals to integrative promise we must also pay attention, I argue, to these broader contextual, pragmatic, or non-epistemic features as well. Understanding Darwin's mixed track record on integrative promise, and thereby helping to resolve the paradox that Endersby raised—why would Darwin have written about orchids instead of people?—requires precisely that we look at the trade-off choices that he made between stringency and opportunism.

[11] Only *starts* to look like such a case, however, because I don't have the space here to talk about why horizontal integration across the tree of life was so important for Darwin's argument at this point in the defense of natural selection, nor its relationship to social and ethical values. I hope the point is reasonable enough to be plausible in the absence of those details.

References

Ankeny RA, Leonelli S (2020) Model organisms. Cambridge University Press, Cambridge

Beatty JH (2006) Chance variation: Darwin on orchids. Philos Sci 73(5):629–641. https://doi.org/10.1086/518332

Beccaloni G (2017) Wallace's moth and Darwin's orchid (unpublished manuscript). https://doi.org/10.13140/RG.2.2.35778.38087

Bellon R (2009) Charles Darwin solves the "riddle of the flower"; or, why don't historians of biology know about the birds and the bees? Hist Sci 47(4):373–406. https://doi.org/10.1177/007327530904700402

Bernhardt P, Edens-Meier R (2014) Darwin's *Orchids* (1862, 1877): origins, development, and impact. In: Edens-Meier R, Bernhardt P (eds) Darwin's orchids: then and now. University of Chicago Press, Chicago, pp 3–19

Boulter M (2008) Darwin's garden: Down House and the origin of species. Constable, London

Conway Morris S (2009) The predictability of evolution: glimpses into a post-Darwinian world. Die Naturwissenschaften 96(11):1313–1337. https://doi.org/10.1007/s00114-009-0607-9

Costa JT (2017) Darwin's backyard: how small experiments led to a big theory. W. W. Norton, New York

Darden L, Maull N (1977) Interfield theories. Philos Sci 44(1):43–64. https://doi.org/10.1086/288723

Darwin C (1839) Letter 523 – Darwin, C. R. to Herbert, William, 26 Jun 1839. https://www.darwinproject.ac.uk/DCP-LETT-523

Darwin C (1859) On the origin of species, 1st edn. John Murray, London

Darwin C (1860a) Fertilisation of British orchids by insect agency. Gardeners' Chron Agric Gazette 23:528

Darwin C (1860b) Letter 2791 – Darwin, C. R. to Henslow, J. S., 8 May [1860]. https://www.darwinproject.ac.uk/DCP-LETT-2791

Darwin C (1862a) Letter 3662 – Darwin, C. R. to Gray, A., 23[–4] Jul [1862]. https://www.darwinproject.ac.uk/DCP-LETT-3662

Darwin C (1862b) On the various contrivances by which British and foreign orchids are fertilised by insects. John Murray, London

Darwin C (1871) The descent of man and selection in relation to sex. John Murray, London

Darwin C (1875) Insectivorous plants. John Murray, London

Darwin C (1876) On the origin of species, 6th edn. John Murray, London

Darwin C (1877) On the various contrivances by which British and foreign orchids are fertilised by insects, 2nd edn. John Murray, London

Darwin C (1881) The formation of vegetable mould, through the action of worms, with observations on their habits. John Murray, London

Darwin F (1887) The life and letters of Charles Darwin, including an autobiographical chapter. John Murray, London

Darwin C (1909) The foundations of the *Origin of Species*: two essays written in 1842 and 1844. Cambridge University Press, Cambridge

Dietrich MR, Ankeny RA, Crowe N, Green S, Leonelli S (2020) How to choose your research organism. Stud Hist Philos Biol Biomed Sci 80:101227. https://doi.org/10.1016/j.shpsc.2019.101227

Edens-Meier R, Bernhardt P (eds) (2014) Darwin's orchids: then and now. University of Chicago Press, Chicago

Endersby J (2016) Orchid: a cultural history. University of Chicago Press, Chicago

Frank SA (2017) Universal expressions of population change by the Price equation: natural selection, information, and maximum entropy production. Ecol Evol 7(10):3381–3396. https://doi.org/10.1002/ece3.2922

Frank SA (2020) Simple unity among the fundamental equations of science. Philos Trans R Soc B: Biol Sci 375(1797):20190351. https://doi.org/10.1098/rstb.2019.0351

Freeman RB (1977) The works of Charles Darwin: an annotated bibliographical handlist, 2nd edn. Archon Books, Dawson

Gould SJ, Lewontin RC (1979) The spandrels of San Marco and the Panglossian paradigm: a critique of the adaptationist programme. Proc R Soc Lond B: Biol Sci 205(1161):581–598

Grafen A (2014) The formal Darwinism project in outline. Biol Philos 29(2):155–174. https://doi.org/10.1007/s10539-013-9414-y

Gray A (1862a) Letter 3637 – Gray, A. to Darwin, C. R., 2–3 Jul 1862. https://www.darwinproject.ac.uk/DCP-LETT-3637

Gray A (1862b) Letter 3736 – Gray, A. to Darwin, C. R., 22 Sep 1862. https://www.darwinproject.ac.uk/DCP-LETT-3736

Green S, Dietrich MR, Leonelli S, Ankeny RA (2018) "Extreme" organisms and the problem of generalization: interpreting the Krogh principle. Hist Philos Life Sci 40(4):65. https://doi.org/10.1007/s40656-018-0231-0

Hedrich R, Fukushima K (2021) On the origin of carnivory: molecular physiology and evolution of plants on an animal diet. Annu Rev Plant Biol 72(1):133–153. https://doi.org/10.1146/annurev-arplant-080620-010429

Krohs U (2012) Convenience experimentation. Stud Hist Philos Biol Biomed Sci 43(1):52–57. https://doi.org/10.1016/j.shpsc.2011.10.005

Leifchild JR (1862) [Review of] On the Various Contrivances by which British and Foreign Orchids are fertilised by Insects, and on the Good Effects of Intercrossing. By Charles Darwin, M.A. Athenaeum 1804:683–685

Lennox JG (2010) The Darwin/Gray correspondence 1857–1869: an intelligent discussion about chance and design. Perspect Sci 18(4):456–479

Levins R (1966) The strategy of model building in population biology. Am Sci 54(4):421–431

Lewontin RC (1970) The units of selection. Annu Rev Ecol Syst 1(1):1–18. https://doi.org/10.1146/annurev.es.01.110170.000245

Mivart SGJ (1871) On the genesis of species, 2nd edn. Macmillan, London

Pence CH (2018) Sir John F. W. Herschel and Charles Darwin: nineteenth-century science and its methodology. HOPOS 8(1):108–140. https://doi.org/10.1086/695719

Pence CH (2021) The causal structure of natural selection. Cambridge University Press, Cambridge

Pence CH (2022) The rise of chance in evolutionary theory: a pompous parade of arithmetic. Academic Press, London

Potochnik A (2017) Idealization and the aims of science. University of Chicago Press, Chicago

Rice SH (2020) Universal rules for the interaction of selection and transmission in evolution. Philos Trans R Soc B: Biol Sci 375(1797):20190353. https://doi.org/10.1098/rstb.2019.0353

Schindler S (2018) Theoretical virtues in science: uncovering reality through theory. Cambridge University Press, Cambridge

Sedgwick A (1860) Objections to Mr. Darwin's theory of the origin of species. Spectator 33:285–286

Shokralla S, Spall JL, Gibson JF, Hajibabaei M (2012) Next-generation sequencing technologies for environmental DNA research. Mol Ecol 21(8):1794–1805. https://doi.org/10.1111/j.1365-294X.2012.05538.x

Sloan PR (2009) The making of a philosophical naturalist. In: Hodge MJS, Radick G (eds) The Cambridge companion to Darwin, 2nd edn. Cambridge University Press, Cambridge, pp 21–43

Wells HG (1894) The flowering of the strange orchid. Pall Mall Budget

Yoccoz NG (2012) The future of environmental DNA in ecology. Mol Ecol 21(8):2031–2038. https://doi.org/10.1111/j.1365-294X.2012.05505.x

Chapter 4
Social Values and Grand Challenges

The Promise of Programmatic Research in Biology

Abstract Another remarkable, recurrent feature of the integrative explanations that I have targeted in this book is their regular appeal to non-epistemic, social, or ethical values. This makes those explanations, and, by extension, integrative promise, a fruitful site for philosophical study of the role of non-epistemic values in science. In this chapter, I explore this connection by looking at the invocation of so-called "grand challenges" in these explanations. Integrative work is often taken to be essential to the resolution of large-scale, difficult social and cultural problems. To illuminate this relationship, I'll consider a recent programmatic document describing future research for the European Molecular Biology Laboratory (EMBL). They consistently link the stakes of the most integrative parts of their program to the resolution of grand challenges, and thus to social and ethical value commitments.

Keywords Non-epistemic values · Social values · Science and values · Grand challenges · European Molecular Biology Laboratory · Planetary biology · Wicked problems

Having looked at the ways in which opportunism and stringency can open the door to the influence of non-epistemic values in the defense of integrative promise, I now want to turn to the most unambiguous manner in which this influence makes itself felt: the relationship between integrative promise and what I will call *grand challenges* in the life sciences.

A wide variety of projects touted for their integrative promise are also very explicitly presented as responding to grand challenges. The Human Genome Project (HGP) was heralded as answering a "grand challenge" from its very inception—a "highly visible and important goal," James Watson wrote early in the program, "similar to the 1961 decision made by President John F. Kennedy to send a man to the moon" (Watson 1990, p. 44). Information about human genome sequences would, he continued,

> provide the ultimate answers to the chemical underpinnings of human existence. They will not only help us understand how we function as healthy human beings, but will also explain, at the chemical level, the role of genetic factors in a multitude of diseases, such

C. H. Pence, *Integrative Promise*, Synthese Library 499,
https://doi.org/10.1007/978-3-031-88471-9_4

> as cancer, Alzheimer's disease, and schizophrenia, that diminish the individual lives of so many millions of people. (Watson 1990, p. 44)

By the end of the HGP around fifteen years later, we were still in the process of tackling these grand challenges, though now they were seen in terms of genomics rather than human genome sequencing. Unmistakeably integrative, this round of grand challenges was "formulated into three major themes—genomics to biology, genomics to health, and genomics to society" (Collins et al. 2003, p. 2). Perhaps somewhat depressingly identical to these 2003 challenges were the series of goals "for improving human health at the forefront of genomics" published by the National Human Genome Research Institute another fifteen years later, this time targeting (even in the same order) "the widespread integration of genomics into almost all areas of biomedical research, the adoption of genomics into mainstream medical and public-health practices, and an increasing relevance of genomics for everyday life" (Green et al. 2020, p. 683). The BRAIN Initiative, a targeted intervention by the Obama administration to support research "that will enable researchers to produce dynamic pictures of the brain that show how individual brain cells and complex neural circuits interact at the speed of thought" was formulated directly as one of "the Administration's 'Grand Challenges'—ambitious but achievable goals that require advances in science and technology" (The White House 2013).

Of course, these are all public-facing, policy-driven projects—it might thus be unsurprising that they invoke grand challenge talk. What we'll see in this chapter (and the next), however, is that this orientation is shared much more broadly across the life sciences than we might have predicted. The more detailed case study of a "grand challenge" narrative which is the focus of this chapter is from a large laboratory in fundamental, molecular biology; in the next chapter (where I'll offer a combined analysis of all the features of integrative promise in a single case), we'll see that the same is true in research in natural history. Grand challenges, then, are a feature of much more than just the science-policy interface.

I'll return below to a few proposals about why integrative promise would be important in the construction of life-science responses to grand challenges, but for now the basic point should, hopefully, be obvious: one repeated aspect of integrative promise seems to be its ability to respond to "grand-challenge" narratives in the biological sciences. Put differently, when biologists want to show that a project might well have grand-challenge-level impact on our society, they often do so in the context of arguments about the integrative promise of their research. As I'll demonstrate, this relationship offers perhaps the most frequent way in which non-epistemic value judgments become tightly linked with integrative promise. An important part of what makes integrative promise a cluster of values worth pursuing is its capacity to speak to these grand challenges, which are in turn very often supported by non-epistemic considerations of the social or ethical value of potential solutions.

4.1 Grand Challenges from Hilbert to Obama

What, exactly, is a grand challenge? For all that I will soon argue that the very idea is tied up with social and ethical value judgments, the concept itself is commonly recognized to come from a very different context: the mathematician David Hilbert's introduction of the series of (in their final, published version) twenty-three open, fundamental problems in mathematics now known as Hilbert's Problems. Positioned, as few scholars of the time were, as someone who "had done his major work in algebra, number theory, and geometry—and, though barely begun, was already doing work in analysis, the fourth major branch of mathematics," Hilbert gave himself the assignment in 1900 "to lecture on the significance of individual problems and to give a list of problems that he thought would be the most fruitful for mathematics in the new century" (Yandell 2002, p. 16).

What made each of these problems relevant for Hilbert, or what elevated a garden-variety question into (what we would now call) a grand challenge, was a kind of balance between tractability, difficulty, and aesthetic virtue. As he wrote,

> Moreover a mathematical problem should be difficult in order to entice us, yet not completely inaccessible, lest it mock at our efforts. It should be to us a guide post on the mazy paths to hidden truths, and ultimately a reminder of our pleasure in the successful solution. (Hilbert 1900, in Yandell 2002, p. 390)

Even a reflection on the list of problems itself is posed in a kind of quasi-mystical fashion. "Who of us," Hilbert asks, "would not be glad to lift the veil behind which the future lies hidden; to cast a glance at the next advances of our science and at the secrets of its development during future centuries?" (Hilbert 1900, in Yandell 2002, p. 389).

This focus on difficult, unsolved problems with paramount social or cultural importance has become something of a preoccupation for scholars across the sciences and humanities. We've just seen that a number of major research programs in the life sciences have presented themselves as unequivocally oriented toward these grand challenges, with precisely the same search for a balance between tractability and significance that we found in Hilbert. Before we continue with an analysis of a putative link between integrative promise and grand challenges, we need to think about just what a grand challenge might actually consist in, and develop some theoretical tools for understanding them.

The discipline that has theorized perhaps most explicitly about the nature of grand challenges and the ways in which we might respond to them is, interestingly, organizational studies. In the last two decades, organizations have been faced with increasing pressure to respond not only to their own local concerns, but to grand challenges themselves. To take an example from my own experience, I imagine that mine is not the only university to have an office dedicated to our response to the climate and biodiversity crisis (or *la transition*, as it is often, somewhat grandiosely, described in French).

It is intuitively clear enough that the magnitude of this sort of grand challenge entirely surpasses the ability of a single institution to "respond" to it in a meaningful way. But how can we make that intuition more precise? What is it that distinguishes grand challenges from other kinds of (more mundane) problems? A helpful taxonomy is provided by Ferraro et al. (2015). As they note, while all stakeholders will agree that responses to grand challenges will be highly diffuse and implicate a vast number of organizations, “we lack a systematic understanding of the mechanisms linking distributed action and the emergence of novel solutions to grand challenges” (Ferraro et al. 2015, p. 364). They propose, then, to make the notion itself more precise by highlighting three characteristics that define those grand challenges. The first is their complexity. Grand challenges involve “many interactions and associations, emergent understandings, and [exhibit] non-linear dynamics” (Ferraro et al. 2015, p. 364). Secondly, such challenges entail what they call “radical uncertainty,” which captures the idea that, when faced with a grand challenge, “actors cannot define the possible future states of the world, and therefore cannot forecast the consequences of their present actions, or whether future others will appreciate them” (Ferraro et al. 2015, p. 364). Third and finally, “grand challenges are evaluative, cutting across jurisdictional boundaries, implicating multiple criteria of worth, and revealing new concerns even as they are being tackled” (Ferraro et al. 2015, p. 364).

We can see that these three features hold even in the case of grand challenges in the otherwise fairly well delimited domain of the life sciences. The kinds of phenomena proposed to be tackled by the HGP involve, as we now know, a bewildering array of factors, from the biochemical, to the genetic, to the environmental, to the social. Any approach that reduces or elides this complexity is, often, extremely problematic, and likely to muddy the waters both scientifically and with regard to the social and political consequences of such research (Morange 2001). Turning to uncertainty, and to take just one example, most predictions about the future utility of human gene sequencing data turned out to be false. This is not to say that the data have not been useful—far from it—but the kind of radical uncertainty described by Ferraro et al. seems to be alive and well in these kinds of scientific cases. And lastly, we also have ample evidence of the evaluative nature of these problems. Put differently, the Human Genome Project was pursued in large part because obtaining the knowledge that it sought would be *a good thing*, and the general sentiment of dissatisfaction surrounding its legacy has much to do with these *normative* failures rather than empirical ones (Lloyd 2008).

Identifying these features of grand challenges, then, lets us draw a connection between this concept and another with a significant literature both within and beyond the philosophy of science: *wicked problems*. This term was originally coined by Horst Rittel and Melvin Webber in 1973, in the context of urban planning and social policy. Why is it, they ask, that professionals are less and less able to handily resolve the large societal problems—the grand challenges—that face society? They respond that, at least in part, this is because (having already handled problems with easier solutions) we dedicate more and more of our time to wicked problems, problems which resist solution in a variety of concrete and specifiable ways. In this paper, they enumerate ten features—which outline a cluster concept, rather than serving as necessary and sufficient conditions—of wicked problems. I'll discuss a few of the most relevant for our purposes here.

Perhaps most importantly of all, wicked problems *do not have a single, definitive formulation*. In part, this is due to their complexity, but more fundamentally, because "the information needed to *understand* the problem depends on one's idea for *solving* it" (Rittel and Webber 1973, p. 161). If the underlying problem of the biodiversity crisis is conceived in terms of habitat loss, then that very conception will govern how we see the solutions (i.e., in terms of the relationship between humans and their environment). And vice versa—if we are already convinced that this is where we will find our solution, then our understanding of the problem will follow along in due course. We might see this as one of the most significant impacts of precisely the kind of radical uncertainty that Ferraro et al. noted above: we are not even able to evaluate future possible solutions and their consequences well enough to offer a univocal statement of the problem at issue.

In part because the formulations of the problems and their solutions are so minimally constrained, it is *difficult*, if not impossible, *to be in possession of a single list of all possible solutions* to such a problem—making a straightforward evaluation of the costs and benefits of all such solutions a non-viable approach to tackling them. Even if we attempt to implement a solution and are evaluating its efficacy retrospectively, this complexity means that underdetermination looms even larger than normal, and we will be able to *ascribe the success or failure of our solution to a large number of different putative causes*. As Rittel and Webber (1973, p. 166) put it, when we evaluate the success of grand-challenge solutions, "the modes of reasoning used in the argument are much richer than those permissible in the sciences. Because of the essential uniqueness of the problem, and lacking opportunity for rigorous experimentation, it is not possible to put [a hypothesis] to a critical test." Consider a concrete example. Should the BRAIN Initiative fail to carry through on its claimed integrative promise, it will not even be clear how to ascribe responsibility for this failure. Were the empirical methods, after all, not fit for purpose? Was the problem badly posed? Or did the failure perhaps lie instead with the normative or societal integration of the empirical results? It will likely be possible to make an argument for any of these explanations, depending on one's interests.

Finally, and the feature that I want to discuss most extensively in this chapter, we have a collection of issues surrounding the *value-ladenness* of solutions to wicked problems. The answers, Rittel and Webber write, to such problems are *not simply to be appraised as true or false*, like the solutions to a mathematical formula. Rather, "assessments of proposed solutions are expressed as 'good' or 'bad' or, more likely, as 'better or worse' or 'satisfying' or 'good enough.'" (Rittel and Webber 1973, p. 163). This normative aspect is then magnified by the fact that we usually have *only one chance to attempt a solution* to the problem in its current form. Since "*every* implemented solution is consequential," and "leaves 'traces' that cannot be undone" (Rittel and Webber 1973, p. 163), any effort to solve this problem will *leave behind it a series of other, different wicked problems in its wake*. As we have already seen, the Human Genome Project seamlessly transitioned into a broad-scale genomics project; successor technologies from epigenetics to the microbiome to organoids are now vying for the right to be the next solution to the wicked problem of personalized medicine.

This thus sets the stage for the two major claims that I want to make in this chapter. In pursuing the analogy between grand challenges and wicked problems, we can explore, on the one hand, why it is that integrative promise might appear so regularly when the life sciences turn their focus to grand challenges. And on the other hand, we will be able to see that such a pursuit gives appeals to integrative promise an inescapably normative dimension. Promise, that is, can give us a particularly attractive way in which to analyze how these value commitments are deployed by life scientists. To that end, I'll explore a case study from the recent mission statement describing the four-year vision of the European Molecular Biology Laboratory (EMBL).

4.2 Promise as a Response to Grand Challenges

To begin, I want to approach the question of the link between integrative promise and grand challenges at a general level. I've gestured at a few examples, like the Human Genome Project, that might intuitively seem integrative and to respond to grand challenges. In this section, I want to offer a more comprehensive argument that, in fact, we should expect these two things to be connected—that is, we should expect to find appeals to the other features of integrative promise that I've already discussed (integration, stringency, and opportunism) when biologists tackle grand challenges.

To see why this might be the case, consider Bryan Norton's reformulation of the central insight of Rittel and Webber (1973):

> Many "problems," as addressed in real-world situations, involve elements that exceed the complexity of any known or hoped-for *model*. No descriptive disciplinary perspective or expert system can embody all of the variables and data necessary to understand, predict, and control the functioning of the dynamic systems within which humans struggle with complex problems and conflicts. (Norton 2012, p. 448)

This is an insight that deserves to be unpacked. Why is this a reasonable way to interpret the much more general framing of wicked planning problems as Rittel and Webber describe them? Often—in the case, again, of environmental problems or climate change—the argument goes by way of the diverging *interests* of the different stakeholders implicated in a wicked problem. If climate change has facets ranging across science, economics, transportation, manufacturing, energy, food, and more, then it is unsurprising that actors will arrive at debates concerning it with not only different understandings of just what the problem *is*, but also different understandings of potential classes of solutions, different values, and different empirical and theoretical backgrounds. Integration in that kind of a context is thus obligatory in a very straightforward sense (the same point is made by Conklin 2006).

Some of the cases of integrative promise that I have discussed and will discuss fall under the scope of this argument. The role of the life sciences in responding to the biodiversity crisis or to climate change will be conditioned by this widespread discord in theoretical perspective, which will therefore need to be taken into account in the practice of science that is relevant for these kinds of policy choices (Hicks et al. 2018). I'll thus come back to this point when I discuss instances of this sort in the case study in the next section.

But not all cases of integrative promise can be described by considering what happens when these concepts move "beyond" the scientific community and encounter diverging epistemic and non-epistemic perspectives. Examples like the Human Genome Project seem to feature a call for vertical integration—from the biochemical through the genetic to the individual and onward to public health—that arises, we might say, "internally" to the scientific community. Perhaps part of the pressure toward personalized medicine came from patients' rights groups or medical professionals, but in general it seems as though the "divergence-of-perspectives" explanation will fail in cases such as these: we are dealing only with a collection of biologists, however diverse that collection might appear.

An argument connecting integration to grand challenges will thus need a second dimension, one that points more to the epistemic features of the kinds of explanations that would need to be deployed in order to confront them. I think such an argument is indeed available. The key point to be made is that when grand-challenge problems reach a certain size, the knowledge needed to be able to tackle them will, almost as a matter of epistemic necessity, become fragmented across multiple actors working in multiple disciplines (Alford and Head 2017, p. 404). The development of personalized medicine on a biochemical or genetic basis, for example, is not only beyond the social-cognitive ability of a single individual to access (this much has been true of central problems in most major scientific disciplines for centuries, as has been stressed in the literature on social epistemology), but lies even beyond the cognitive ability of a single disciplinary perspective to access.

We thus can see a way in which, even if every actor involved in the solution of a grand challenge might broadly describe themselves as working in the life sciences—and, therefore, the thornier problem of radically incommensurable perspectives on the problem is likely absent—we still have an epistemic demand for integration. The examples I've discussed so far have largely been about vertical integration, but we can imagine analogous cases for horizontal integration as well: demonstrating that evolutionary theory can genuinely hold across the tree of life was, even in Darwin's day, the work of natural historians, paleontologists, zoologists, botanists, and more.

An even simpler version of the argument might approach the question of the grand challenge–integrative promise link from the other direction. Just what we mean when we describe a problem as "wicked" is to say that it has resisted traditional efforts at solution of the sort that would be pursued within the confines of traditional disciplines approaching problems with their traditional tools. Wicked problems are what's *left* after we eliminate all of the problems that don't require integration.

I think there are very good reasons, then, to think that "grand challenge" type theorizing bears a very deep connection to the pursuit of integration. The same is also the case for stringency and opportunism. The opportunistic nature of grand-challenge solutions is perhaps most strongly cemented by the value-laden nature of these problems, which guarantees that they have a kind of ineliminable urgency. We want to solve them not in the abstract, but in the here and now, because they pose serious and genuine epistemic and social dilemmas. That requires that we get to work, and, by extension, that we get to work with the tools that we have at our disposal. At the same time, as we saw in the last chapter, the options for resolving them will be limited, and they will need to meet stringent standards. Both of these latter features will, then, leave important pragmatic marks on our efforts to find solutions to grand challenges, and will shape the landscape of solutions that we propose—it will be necessary to keep them in mind, as I've already proposed, to fully understand which solutions are promoted and why.

It is paramount, however, not to let the framework of wicked problems become too dominant in our thinking, nor to claim that wicked problems are a precisely delimited category. As John Alford and Brian Head (2017) have argued, the two central features of wicked problems, the fragmentation of knowledge that I just discussed, and the complexity (or lack of clear formulation) of the problem and solution that I mentioned a bit earlier, are both features that come in degrees. A strict, binary classification of problems into "wicked" and "tame" is therefore likely too quick, and will smooth over important distinctions in our potential responses. That said, we can keep this warning in mind without threatening the argument that I've advanced here. In short, the more wickedness seems to be present in a particular case the more likely that, on the one hand, it will be labelled a "grand challenge," and, on the other, that solutions to it will be touted for their integrative promise. Just as whether a problem is wicked or not comes in degrees, the pattern of epistemic and non-epistemic values that I've dubbed integrative promise can be exemplified in degrees as well.

4.3 The Value-Ladenness of Wicked Problems

So much for the first claim that I want to defend in this chapter: that the kinds of problems highlighted by grand-challenge rhetoric are likely to be wicked. My second, and more important, goal is to use that claim to draw a further inference about grand-challenge problems: that they are often inescapably tied to non-epistemic value judgments. I thus want to argue that the link between integrative promise and grand challenges constitutes yet another way in which non-epistemic value judgments become frequently and crucially implicated in evaluations of integrative promise. I'll start by making the case more generally, and then close the chapter with a concrete case study.

In their original presentation of wicked problems—recall, in the much more obviously ethically and politically charged context of planning and political decision-making—Rittel and Webber noted that

> Our point ... is that diverse values are held by different groups of individuals—that what satisfies one may be abhorrent to another, that what comprises problem-solution for one is problem-generation for another. Under such circumstances, and in the absence of an overriding social theory or an overriding social ethic, there is no gainsaying which group is right and which should have its ends served. (Rittel and Webber 1973, p. 169)

In short, insofar as one frequent cause of wickedness is differences in social or ethical value judgments, efforts to resolve wicked problems will very often founder on debates over non-epistemic value commitments. This is a feature of wicked problems often repeated throughout the literature. Alford and Head introduce as one of their criteria for wickedness the possibility that "the various stakeholders have interests (or values) which are substantially in conflict with those of others" (Alford and Head 2017, p. 407). Jeffrey Conklin notes that judgments of the 'better' and 'worse' quality of solutions "are likely to vary widely and depend on the stakeholder's independent values and goals" (Conklin 2006, p. 8). In exploring their case study of climate change, Ferraro et al. note that "the very definition of the problem and the actors assembled around its resolution are value-laden issues" (Ferraro et al. 2015, p. 367).

Examples could be multiplied, but the basic point remains. Wicked problems do not admit of a straightforward enumeration of all their possible solutions. Because of this, they equally don't admit of simplistic, complete cost-benefit analyses of our potential choices. In turn, solutions will and must be evaluated comparatively, as meeting or failing to meet minimal requirements. Any such evaluative regime will be radically underdetermined by the available empirical data. As has been compellingly argued across the last thirty years of the philosophy of science and values literature (Douglas 2000; Biddle 2013; Douglas 2021; though see Brown 2013), cases of such underdetermination, even if only a temporary condition of current scientific knowledge, are often filled in by the non-epistemic value commitments of the scientists involved.

And this is only half of the argument. It is also relatively clear that, at least in many cases, grand challenge problems are chosen precisely *because* of their value-laden consequences. We pursue something like the BRAIN Initiative not only because of the intrinsic value of the knowledge that it can offer us, but also because it "ultimately aims to help researchers find new ways to treat, cure, and even prevent brain disorders, such as Alzheimer's disease, epilepsy, and traumatic brain injury" (The White House 2013). The "ultimately" here, I think, is no rhetorical flourish: the genuine, long-term aim of the project is fundamentally tied to the fact that treating brain disorders amounts to a moral good.

Even in cases of what might appear to be more "strictly scientific" pursuits, recent work in HPS has emphasized the extent to which these kinds of non-epistemic concerns are difficult to eliminate. To take only a few examples, Marga Vicedo (1995) has noted that the early development of Mendelian genetics is intrinsically linked to agricultural efforts of the day, Jon Hodge (2009) has pushed the same point for Darwin's pursuit of natural selection, and Staffan Müller-Wille and Hans-Jörg Rheinberger (2012) have advanced a similar argument for rethinking the appropriate tradition for reading Mendel. In a more contemporary context, Hannah Landecker (2007) has advocated a pragmatic and value-laden interpretation of the study of cell culture (and, by extension, the domains of molecular biology that rely on it). The development of these kinds of vast, integrative domains of knowledge—from evolutionary theory to cellular biology—is almost always undertaken with a central role for the social and ethical value consequences of that research lying in the background.

All told, there are good reasons to think—whether directly or indirectly—that grand challenges will almost necessarily invoke non-epistemic value judgments. If they do so, and integrative promise is deeply connected with the pursuit of grand challenges, then it will be important to recognize and explore these non-epistemic value judgments in cases of integrative promise. This much I think we can demonstrate at the abstract level: let's turn to a concrete case study to see how this looks in practice.

4.3.1 A Program for the Future: EMBL

The European Molecular Biology Laboratory (EMBL) was founded in 1974 as a home for life-sciences research analogous to the fundamental physics undertaken in the European Union by CERN. As Bruno Strasser has noted, in the environment of Cold-War domination of the scientific research agenda by atomic and space projects, "it was far from clear … that molecular biology deserved a comparable level of governmental support, over and above other areas of the biomedical sciences" (Strasser 2003, p. 540). After an initial era of strengthening of national molecular biology competences, the next logical step was the founding of a central, international laboratory, which would become EMBL (Rheinberger 2012, p. 641).[1]

Today, EMBL spans six sites and is linked to scientific efforts in dozens of countries. It also offers an interesting opportunity for philosophical study, in that it explicitly publishes its mission statement, updated every five years as EMBL renews what it calls its "programmatic funding" from its member states (European Molecular Biology Laboratory 2022b, p. 7). The latest edition, developed over 2019 and 2020, describes the organization's orientation for the period from 2022 until 2026.[2]

---

[1] The detailed history of the founding of EMBL has yet to be told, as Strasser (2003) has noted. An initial discussion of the complex environment of European molecular biology in the 1960s and 1970s has been developed by John Krige (2002).

[2] The report is published as an executive summary (European Molecular Biology Laboratory 2022b), which also appears as the initial section of the complete version (European Molecular Biology Laboratory 2022a). I'll cite the short version when possible, as it's easier to navigate.

The overarching structure of the report follows EMBL's five missions: research, scientific services, training, technology transfer, and coordination of life science in Europe. The research mission, in turn, is split into six focal areas, supplemented by two transversal or support fields (data science and biological theory). From the program's name—Molecules to Ecosystems—straight through each of the seven particular research foci, the project is loaded with appeals to the integrative promise of the research that EMBL will pursue.

At the broadest level, its mission targets the development of biological understanding "from the molecular building blocks of life through to the complexity of ecosystems," with an ambition "to establish the molecular basis of life in context, to gain new knowledge that is relevant to understanding life on Earth, and to provide translational potential to support advances in human and planetary health" (European Molecular Biology Laboratory 2022b, p. 2). What has made this program possible, in turn, is recent "incredible advances in technology and quantitative data generation," which have given biologists "the capacity to follow the dynamics of living matter in real time and at multiple scales" (European Molecular Biology Laboratory 2022b, p. 3). From the very beginning, and even when the project is presented in the broadest terms, we see a commitment to *vertical integration*, with the explicit goal to build a unified picture of biology from the molecular to the ecosystemic; to *stringency*, as this goal is constantly and carefully nuanced to pick out cases important to generating integrated knowledge; to *opportunism*, as developments are constantly positioned with respect to EMBL's current interests; and to *grand challenges*, as the proposed research is offered at "a critical time for society," responding to "the spread of infectious diseases, the loss of biodiversity, environmental degradation, and climate change" (European Molecular Biology Laboratory 2022b, p. 3). In short, the programmatic presentation of EMBL's research direction as a whole very precisely exemplifies integrative promise.

The same carries through each of the six more detailed research foci: molecular building blocks in context, cellular and multicellular dynamics of life, microbial ecosystems, infection biology, human ecosystems, and planetary biology. The first research focus targets understanding the development of phenotypes at levels from the molecular to the environmental, leveraging "EMBL's past and current research and strengths in technology development" to "investigate biological complexity in a way that takes account of environmental variation at the subcellular and molecular levels" (European Molecular Biology Laboratory 2022a, p. 28), clearly invoking both vertical integration and opportunism. "Although each scale is being actively studied," they write, "scales are typically studied in isolation with little integration across them" (European Molecular Biology Laboratory 2022a, p. 31), thus motivating EMBL's integrative approach. The same vertical integration powers the second focus on cellular and molecular dynamics of life. An understudied dimension of this dynamics, they write, "is the way in which the different levels of organisation in living systems are interlinked, influencing and feeding back on each other in a reciprocal manner" (European Molecular Biology Laboratory 2022a, pp. 44–45); only via vertical integration can we understand how this feedback gives rise to the crucial interplay of robustness and plasticity.

The third research area, into microbial ecosystems, employs EMBL's extant work on human microbiome research, now applying it to the much less studied, more stringent domain of non-human microbial communities. Here, the main integration target is horizontal. Many species in the human microbiome are of unknown functional significance (this is repeatedly referred to in the mission document, with a nod to theoretical physics, as the "dark genetic matter" in microbial communities), and the program includes concrete plans to "isolate, biobank, and sequence thousands of strains from human individuals to build a repository for functional studies and for understanding the dynamics and properties of different microbial ecosystems" (European Molecular Biology Laboratory 2022a, p. 61). The infection biology research arm, similarly, targets "the diversity of pathogens, which include viruses, bacteria, fungi, protozoa, and other parasites. Within each of these groups, there is staggering diversity, both in their fundamental biology and ecology" (European Molecular Biology Laboratory 2022a, p. 73). Here it is biology's lack of horizontal integration across the diversity of microbial and pathogenic systems that impedes science's ability to rise to meet the grand challenges at issue.

The fifth research area, on human ecosystems, returns to vertical integration, seeking to integrate understanding of "physical, biological, and social environments" with an eye toward "how environmental factors can precipitate disease, and more generally, how genotype and the environment influence human phenotypes" (European Molecular Biology Laboratory 2022a, p. 91). The reason to pursue this research directly invokes opportunism and grand challenges: this line of study should be pursued because, on the one hand, "environmental challenges are increasingly impacting human health," and on the other, "science is experiencing an exponential rise in the data available on human molecular and related environmental measurements" (European Molecular Biology Laboratory 2022a, p. 93). The final research thread, planetary biology (which will be the topic of its own more detailed analysis in the next subsection), directly appeals to the collection of "new information at the molecular, cellular, and organismal levels" to describe ecosystem function in the context of the biodiversity crisis (European Molecular Biology Laboratory 2022a, pp. 107–108).

We thus see that the commitment to integration (both horizontal and vertical), stringency, and opportunism runs deep throughout every major piece of the research mission in the new EMBL program. Emphasizing the approach in terms of integrative promise lets us see how these facets interact. For instance, as we also saw in the last chapter, there is a complex interplay between opportunism and stringency in each of these research themes. On the one hand, EMBL's researchers already pursue a huge number of projects, and in particular have developed expertise with a variety of technologies. But those extant research projects are insufficient to develop the kind of horizontal and vertical integration that they argue will be necessary to approach the problems they're interested in solving. This tension between the opportunistic availability of current research projects and the stringency set in place by the difficulty of work as ambitious as that proposed here is another recurring theme throughout the document, and it is precisely one of the characteristics for which we would search in analyzing their program through the lens of integrative promise.

To narrow in on the theme of this chapter, then, each of these research initiatives also is phrased straightforwardly in terms of their ability to respond to a variety of grand challenges. For the first focus, studying molecular building blocks in context will "pave the way for precision interventions to mitigate the harmful effects of our ever-changing world on living systems, including humans, and will help inform policymakers on issues relating to planetary and human health" (European Molecular Biology Laboratory 2022a, pp. 42–43). The second research focus will have not only impacts for healthcare, but also in "revealing how living systems can deteriorate in ageing, disease, or through disruptive environmental changes" (European Molecular Biology Laboratory 2022a, p. 56). Work on microbial ecosystems will push microbiome-based personalized medicine, as well as "unprecedented opportunities for reviving and rebalancing natural ecosystems" (European Molecular Biology Laboratory 2022a, p. 58). Infection biology, particularly in the context of COVID, has relatively evident social- and ethical-value valence, but the crisis of anti-microbial resistant pathogens is particularly emphasized.

Human ecosystems research is targeted at the impacts of environmental factors on disease and public health outcomes, in line with the broader societal focus "on mitigating exposure to detrimental influences such as dietary factors, stress, pollutants, and infectious pathogens" (European Molecular Biology Laboratory 2022a, p. 105). The planetary biology area, finally, openly argues for its potential to develop "solutions to pressing environmental and societal challenges, including global warming, environmental pollution, harmful zoonotic diseases, and AMR [anti-microbial resistance]" (European Molecular Biology Laboratory 2022a, p. 109).

There are a whole host of arguments expressing non-epistemic value commitments deployed across the document, whether explicitly or implicitly, and there are a variety of ways in which we might analyze them. But we can notice some commonalities, which seem to represent something like an institutional commitment to a set of non-epistemic values that runs throughout these evaluations of the potential impact of EMBL's proposed research. EMBL positions its mission as a response to *global crises*—declining human health, anti-microbial resistant pathogens, the loss of biodiversity, climate change, and global pandemic disease foremost among them. Reaction to these crises will include, at least in part, the development of solutions from molecular biology, where those solutions require vertical and horizontal integration of a sort that is not currently available. The institution thus positions its technical program not only as one that will advance the state of global knowledge in molecular biology, but also as one which will make the world a better place in well-defined ways, by responding to this collection of large-scale grand challenges.

4.3.2 Deep Dive: Planetary Biology

> *Planetary biology research represents an extraordinary fusion between intrinsically fascinating biological questions and modern environmental challenges, which EMBL can and must help to overcome.* (European Molecular Biology Laboratory 2022a, p. 125)

Let's unpack one of these research strands in a bit more detail, to try to understand its emphasis on integration, its invocations of opportunism and stringency, and its relationship to grand challenges and non-epistemic value judgments. To that end, let's explore the last, and arguably most difficult, research theme: what EMBL calls "planetary biology."

The best way to describe the epistemic goals of the problem is explicitly in terms of integration. The intention is "to understand life in its natural context," by which they mean relationships between organisms and their environments. But not just at the traditional neighboring scales of individuals, populations, and ecosystems: "Exploring these relationships *in molecular terms at the ecosystem scale* represents an even greater challenge, yet this is the context in which life happens" (European Molecular Biology Laboratory 2022a, p. 107). The project thus specifically seeks to integrate data from "the molecular, cellular, and organismal levels": as they argue, how can we fully understand the impact of the microbiome on nutrition, for instance, without tracing microbial dynamics in naturally occurring environments (including microbiome or holobiont contexts), the processes that control the development of those communities over time (genetic, epigenetic, and environmental), their interactions with the immune system and their organism-level effects, and the way that external environmental features impact all of the above?

This focus on combining molecular, cellular, organismic, and environmental theories and data in order to understand large-scale scientific problems makes vertical integration a central part of the planetary biology research project (see a graphical representation in Fig. 4.1).[3] The lack of vertical integration in current research is positioned as one of the key advantages of the new program. To quote at length:

> Today, molecular ecology—the application of molecular techniques to ecosystems research—focuses mainly on the use of metagenomic information..., metatranscriptomics..., and barcoding ... to identify species. Although metagenomics provides repertoires of genes and species within ecosystems, it cannot provide multidimensional spatio-temporal information about how organisms coexist (in communities, as symbionts, etc.) or how they change their molecular and cellular phenotypes in different environmental conditions, be they nutritional variations, natural seasonal changes, or environmental insults such as abnormal temperature or the presence of pathogens or chemical toxins. (European Molecular Biology Laboratory 2022a, pp. 111–112)

That is, while current work in molecular ecology can provide a kind of *horizontal* integration, expanding our understanding of the diversity of the species present in

---

[3] They also make explicit reference to integration across time-scales, from "cellular/molecular time" through "seasonal variation" to "multiple years," another kind of scale integration that we discussed in Chap. 2.

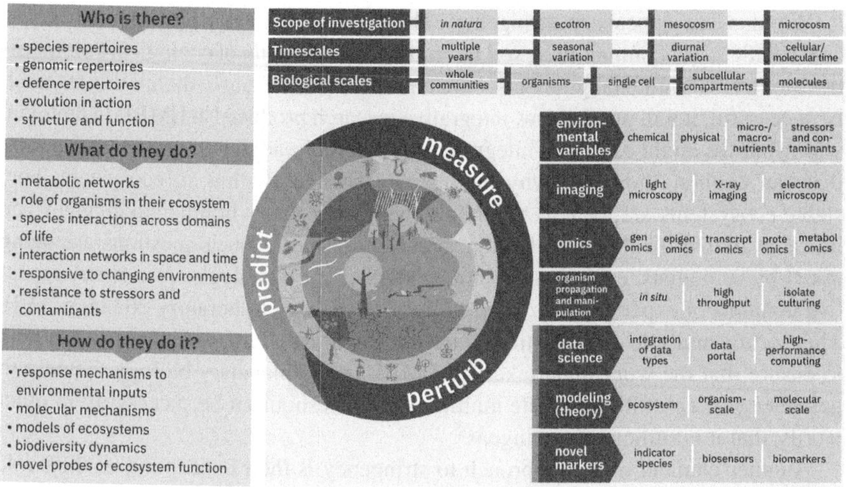

Fig. 4.1 A graphical representation of the integrative scope of EMBL's "planetary biology" research program. Figure PB3 from European Molecular Biology Laboratory (2022a, p. 111), CC-BY 4.0

a given ecological context, it is the absence of *vertical* integration—connections between molecular and cellular behavior and environmental conditions, or between changes in environment and later changes at the micro-level—that really makes the need for a planetary biology program acute.

The project is also fundamentally opportunistic, as it is presented as relying on a convergence of two branches of recent work at EMBL. First are big-data and data-science approaches in molecular biology, which (when combined with data-producing tools like sequencing and gene editing) means "that biologists can now effectively study life in its natural context" (European Molecular Biology Laboratory 2022a, p. 109). And second are a number of extant collaborations between EMBL and horizontally-integrative projects designed to produce broad surveys of ecological diversity, particularly of soil microbes and ocean microbes. The Tara Oceans project, with which EMBL heavily collaborated, was a four-year, open-ocean sampling effort, which sought "discoveries about the diversity of viral, prokaryotic, and eukaryotic species, their genetics, and their interactions, including novel symbiotic relationships" (European Molecular Biology Laboratory 2022a, p. 109)—a classic horizontal-integration goal. This positions EMBL well to synthesize this horizontal and vertical integration work in pursuit of the goals of the planetary biology program.[4]

[4] *Failures* of opportunism are also pointed out. One pilot project will require expertise in botany, and the document notes that "although no plant scientists currently work at EMBL, new group leaders in this area may be recruited in the future" (European Molecular Biology Laboratory 2022a, p. 120).

We also see appeals to stringency. The three initial pilot projects selected—coastal microbial communities, soil microbes, and sampling of ecological communities along environmental gradients—are chosen in part because, should this research be successful, it will put the new, integrative research produced at EMBL in dialogue with what we might call "non-integrative" projects already in progress. Information that can explain "how and why biodiversity changes within an ecosystem, how stable ecosystems are, or how reversible the impact of environmental changes can be" is available for direct evaluation, precisely because "such questions are being asked by ecologists, evolutionary biologists, and epidemiologists," in their respective domains of expertise (European Molecular Biology Laboratory 2022a, p. 112). This sort of multiple confirmation (what we might call "cross-checking") can help to ensure that the work undertaken by EMBL in the planetary biology theme will genuinely, successfully integrate multiple domains should it be successful: in other words, that it is sufficiently stringent.[5]

Another element of this approach to stringency is their conception of their role as a facilitator for the future of molecular biology research in Europe. EMBL wants not only to produce a program which will serve as a stringent test of their own theoretical commitments—they want to play that role across Europe and for years to come:

> EMBL's aim is not to explore all ecosystems, but to focus on dedicated collaborative projects that will serve as **paradigms for future endeavours to study ecosystems at the molecular level.** The intention is not to become an ecological research organisation, but rather to collaborate and build bridges with communities of scientists to build a new era of European life sciences together. (European Molecular Biology Laboratory 2022a, p. 110, original emphasis)

Again, this draws our attention to the careful process of selection of pilot projects that they have deployed, situated between the twin demands of opportunism and extant research in process at EMBL and the desire to produce stringent tests of integrative ecological research that could serve as a model for the coming decades of study in molecular biology.

Given that the planetary biology theme seems to exhibit all of the other characteristics of integrative promise, let's return to the main focus of this chapter: non-epistemic values and grand challenges. Again, as I've already argued above, EMBL's primary set of non-epistemic value commitments seem to be to a vision of science as useful to respond to a world in crisis. Here, the primary crises are biodiversity and climate change, with secondary benefits for human beings. In general,

> ecosystem function is crucial for human survival (for example, for food production, medicine, consumer goods, and materials). However, as biodiversity declines at alarming rates due to habitat destruction, deforestation, urbanisation, pollution, climate change, and

[5] They also appeal to stringency in the effort to construct novel model organisms that can, they hope, more fully engage the integrative questions that are at the heart of the program; for more on this connection between model organisms and integrative promise, see Chap. 7.

> other anthropogenic effects, natural ecosystems are changing on a massive scale. (European Molecular Biology Laboratory 2022a, p. 108)

The character of the threat is, of course, not surprising. What makes their approach different—what constitutes the real link between the non-epistemic values and their epistemic project—comes just at the end of that quote. While we can have some idea of what kinds of anthropogenic influences might be impacting a given ecosystem, as a rule, we do not know what those influences will actually *do* to ecosystems in nature. How will a finely balanced ecosystem respond to an increase in temperature, a loss of a keystone predator, or a change in abiotic resources? In some isolated cases we have answers to these questions, but we lack a vertically integrative story that can help guide research into those reactions.

The potential impact of this ability to develop explanations that span molecular, cellular, organismic, and ecosystemic responses to environmental change should not, they argue, be underestimated. As already quoted earlier, they argue that

> The knowledge gained from EMBL's planetary biology research will enable scientists to answer fascinating fundamental questions about the impact of the environment on biological processes, and will pave the way for solutions to pressing environmental and societal challenges, including global warming, environmental pollution, harmful zoonotic diseases, and [anti-microbial resistance]. (European Molecular Biology Laboratory 2022a, p. 109)

Again, the key link is between processes of *environmental change* and human-scale grand challenges. As is often noted in the literature on wicked problems, one of their most distinctive features is the fact that it is often very difficult to find commonalities between them—as Rittel and Webber put it, wicked problems are "essentially unique," by which they mean that "there are no *classes* of wicked problems in the sense that principles of solution can be developed to fit *all* members of a class" (Rittel and Webber 1973, p. 164). But this is not to say that it is impossible that we might discover certain shared features, connections, or useful cross-cutting tools to aid us in our hunt for solutions. Rittel and Webber are surely correct that "one can never be *certain* that the particulars of a problem do not override its commonalities" (Rittel and Webber 1973, p. 165)—but I don't think EMBL's approach demands such a high level of certainty. The mere possibility of a common approach, grounded in the integrative understanding of environmental change, and useful to such a collection of important grand challenges, goes a long way toward justifying this type of research.

The nature of EMBL's non-epistemic value commitments thus begins to come into focus. The plethora of planetary-scale crises with which we are all now intimately familiar—climate change, biodiversity loss, anti-microbial resistance, pandemic disease—gives us compelling ethical justification to undertake work which could respond to those challenges. Particularly valuable would be any possibility of commonality or unified solution between those otherwise relatively disparate challenges. EMBL believes that they've lighted on just such a shared feature: molecular-to-ecosystemic integrative understanding of responses to environmental change. This, then, serves as the key link between their non-epistemic and epistemic value commitments, and, in particular, as a vital justification for the existence of the project as a whole. Grand-challenge appeals thus help to underwrite the pursuit of what are otherwise, as we have already repeatedly seen, extremely daunting research projects.

An important objection should be treated here before we move on from this case study. One might see the kind of document that I've analyzed here as "beneath" philosophical study. Grant proposals of this sort are, as the objection would have it, little more than hype, with no epistemic content worthy of note and not to be taken seriously as programmatic direction for scientific research. As soon as the ink is dry on the mission statement, an especially cynical take might have it, the scientists at EMBL will abandon all this talk of integration and grand challenges—will abandon integrative promise—and go back to performing the same kind of siloed, mono-disciplinary work that characterizes so much of the rest of the sciences.

Less skeptically, but still posing an important difficulty for my position, Sabina Leonelli has argued that the pursuit of grand challenges might lead to science that is less important *qua* contribution to the sciences. In her words, scientists

> are investing a significant amount of their resources on producing results that might not be revolutionary in terms of their conceptual contribution to existing biological knowledge (though they may prove to be such!), but rather are primarily meant to serve a wider social agenda. (Leonelli 2016, p. 158)

Rather than being an epistemic virtue that works in concert with other parts of integrative promise like inter-domain integration and stringency, then, the pursuit of grand challenges might drive research away from other promising advances—perhaps for very good reasons, but nonetheless fitting awkwardly at best in the context of my story here.

In some sense, there is no answer to these kinds of objections: of course, it is entirely possible that the pursuit of grand challenges is either hollow or, ultimately, counterproductive. But I don't think it would be fair to these scientists to dismiss the entire approach either as pure bluster, designed only to persuade a funding agency, or orthogonal from their pursuit of pure biological knowledge. The research questions that they present are reasonable, and the integrative promise of the research program is, in the end, well argued. In short, integrative knowledge of the kind discussed here not implausibly *would* aid in our response to the environmental grand challenges that they highlight, and it not implausibly would equally advance the biological state of the art. It forms part of an epistemic defense of the proposed research that it behooves us as philosophers of science to take seriously.

On the contrary, I think that cases like this show us the importance of the integrative promise framework for analyzing documents like EMBL's program. The elements of integrative promise aren't simply scattered throughout the document like a checklist—they are tightly woven parts of the argument for the pursuit of the planetary biology project. Approaching them as a unit—approaching them *as* integrative promise—and seeing how those parts interact lets us better understand the epistemic content of their work (why they've chosen the experiments that they have) as well as the non-epistemic position of their organization (what value commitments they believe those experiments will advance). The whole, here, is more than the sum of its parts, as I have claimed will often be the case when studying integrative promise.

At this point, we have seen all of the major features which, I argue, distinguish the pattern of integrative promise: integration, stringency, opportunism, and engagement with non-epistemic values. In that sense, EMBL's planetary biology program is the first "fully fledged" case study that I've offered so far of the integrative promise of a class of scientific explanations. But our focus in this chapter was still largely trained on the question of grand challenges and their attendant non-epistemic value commitments. It's time we turned to a detailed case study—not just of one of the aspects of integrative promise but, I argue, a case in which an active and heated debate in the scientific community can be well understood as a debate over whether or not a particular kind of scientific work exhibits integrative promise in its entirety.

References

Alford J, Head BW (2017) Wicked and less wicked problems: a typology and a contingency framework. Policy Soc 36(3):397–413. https://doi.org/10.1080/14494035.2017.1361634

Biddle J (2013) State of the field: transient underdetermination and values in science. Stud Hist Philos Sci 44(1):124–133. https://doi.org/10.1016/j.shpsa.2012.09.003

Brown MJ (2013) Values in science beyond underdetermination and inductive risk. Philos Sci 80(5):829–839. https://doi.org/10.1086/673720

Collins FS, Green ED, Guttmacher AE, Guyer MS (2003) A vision for the future of genomics research. Nature 422(6934):835–847. https://doi.org/10.1038/nature01626

Conklin J (2006) Wicked problems and social complexity. In: Dialogue mapping: building shared understanding of wicked problems. Wiley, Chichester, pp 3–40

Douglas HE (2000) Inductive risk and values in science. Philos Sci 67(4):559–579

Douglas HE (2021) The rightful place of science: science, values and democracy. The 2016 Descartes Lectures. Consortium for Science, Policy & Outcomes, Tempe

European Molecular Biology Laboratory (2022a) Molecules to ecosystems: the EMBL Programme 2022–2026. Tech. rep. EMBL, Heidelberg. https://www.embl.org/about/programme/

European Molecular Biology Laboratory (2022b) Molecules to ecosystems: the EMBL Programme 2022–2026. Executive summary. Tech. rep. EMBL, Heidelberg. https://www.embl.org/about/programme/

Ferraro F, Etzion D, Gehman J (2015) Tackling grand challenges pragmatically: robust action revisited. Org Stud 36(3):363–390. https://doi.org/10.1177/0170840614563742

Green ED, Gunter C, Biesecker LG, Di Francesco V, Easter CL, Feingold EA, Felsenfeld AL, Kaufman DJ, Ostrander EA, Pavan WJ, Phillippy AM, Wise AL, Dayal JG, Kish BJ, Mandich A, Wellington CR, Wetterstrand KA, Bates SA, Leja D, Vasquez S, Gahl WA, Graham BJ, Kastner DL, Liu P, Rodriguez LL, Solomon BD, Bonham VL, Brody LC, Hutter CM, Manolio TA (2020) Strategic vision for improving human health at the forefront of genomics. Nature 586(7831):683–692. https://doi.org/10.1038/s41586-020-2817-4

Hicks DJ, Stahmer C, Smith M (2018) Impacting capabilities: a conceptual framework for the social value of research. Front Res Metr Anal 3. https://doi.org/10.3389/frma.2018.00024

Hodge MJS (2009) Capitalist contexts for Darwinian theory: land, finance, industry and empire. J Hist Biol 42(3):399–416. https://doi.org/10.1007/s10739-009-9187-y

Krige J (2002) The birth of EMBO and the difficult road to EMBL. Stud Hist Philos Biol Biomed Sci 33(3):547–564. https://doi.org/10.1016/S1369-8486(02)00013-4

Landecker H (2007) Culturing life: How cells became technologies. Harvard University Press, Cambridge

Leonelli S (2016) Data-centric biology: a philosophical study. University of Chicago Press, Chicago
Lloyd EA (2008) Normality and variation: the Human Genome Project and the ideal type. In: Science, evolution, and politics. Cambridge University Press, Cambridge. https://doi.org/10.1017/CBO9780511619724
Morange M (2001) L'eugénisme aujourd'hui. In: Iacub M, Jouannet P (eds) Juger la vie : les choix médicaux en matière de procréation. Éditions La Découverte, Paris, pp 16–34
Müller-Wille S, Rheinberger HJ (2012) A cultural history of heredity. University of Chicago Press, Chicago
Norton BG (2012) The ways of wickedness: analyzing messiness with messy tools. J Agric Environ Ethics 25(4):447–465. https://doi.org/10.1007/s10806-011-9333-3
Rheinberger HJ (2012) Internationalism and the history of molecular biology. In: Renn J (ed) The globalization of knowledge in history. Max Planck Institute for the History of Science, Berlin, pp 637–643
Rittel HWJ, Webber MM (1973) Dilemmas in a general theory of planning. Policy Sci 4:155–169
Strasser B (2003) The transformation of the biological sciences in post-war Europe. EMBO Rep 4(6):540–543. https://doi.org/10.1038/sj.embor.embor879
The White House (2013) Fact Sheet: BRAIN Initiative. https://obamawhitehouse.archives.gov/the-press-office/2013/04/02/fact-sheet-brain-initiative
Vicedo M (1995) What is that thing called Mendelian genetics? Soc Stud Sci 25(2):370–382
Watson JD (1990) The Human Genome Project: past, present, and future. Science 248(4951):44–49
Yandell BH (2002) The honors class: Hilbert's problems and their solvers. A. K. Peters, Natick

Chapter 5
Integrative Promise in Action

The Case of Natural History

Abstract This chapter closes the first part of the book by presenting the first thorough account of the way in which a single case study exemplifies all of the elements which constitute integrative promise. Much recent biological literature has strenuously argued for the utility of traditional, observational natural history, especially as confronted with contemporary molecular phylogenetic methods. I argue here that the best way to understand the state of play around this question is to see it as a disagreement over the actual or potential integrative promise of natural history. According to some, natural history is key for integrating extant bodies of biological theory in response to the global biodiversity crisis; for others, different parts of biology or ecology will play this role. Integrative promise thus helps us understand numerous features of an otherwise somewhat perplexing "debate."

Keywords Natural history · Prediction · Generalization · Biodiversity · Integration · Stamp collecting

We've now seen all of the elements which, I argue, comprise integrative promise. Explanations that are prized for their integrative promise demonstrate inter-domain integration (often either vertical, across spatial and temporal scales, or horizontal, across clades in the tree of life), stringency, and opportunism, and invoke non-epistemic values, especially in their contribution to the resolution of grand challenges. We saw all of these elements come together in the last chapter when we analyzed the programmatic mission of the European Molecular Biology Laboratory. But our discussion there was still largely centered on the non-epistemic value component of promise. In this chapter, we'll take a step back, and, exploring all of these dimensions at the same time, consider a case that, as we will see, directly involves scientists' evaluation of the integrative promise of a line of research.

Analyses of the history of biology often note that Darwin is an interesting sort of hybrid figure, concerned both with constructing overarching theories that explain broad patterns in the living world and with fine-grained explanations of local features of natural populations (like the relationships between cats, mice, and bumblebees described in the *Origin*). Discord over the appropriate role for this

C. H. Pence, *Integrative Promise*, Synthese Library 499,
https://doi.org/10.1007/978-3-031-88471-9_5

latter kind of "organism-centric" biology has been a hallmark of meta-level conflict about the direction of the life sciences relatively constantly since Darwin's day. The first notable such friction formed a part of the stakes during the introduction of mathematical methods into evolutionary theorizing in what is now known as the "biometry-Mendelism" debate (Pence 2022). Morphological explanations grounded in painstaking observation of individual organisms had been the order of the day for hundreds of years, and scholars like William Bateson saw no reason to replace them with abstracted mathematical models of population change.

Whether ecology would remain allied with a kind of observational approach to the living world or would also become mathematized became a heated subject throughout the mid-twentieth century, implicating not only narrower questions about the role of observation, but broader worries about the relationship between ecology and evolutionary theory as well (Kingsland 1995). A similar collection of concerns, though grounded in a different threat to "organism-centric" study, grew out of the development and rapid advance of molecular-biological methods in the life sciences. Fears abounded that a classic vision of science as hypothesis testing would be left behind in the "deluge" of now readily available and reasonably inexpensive gene-sequencing data. The point has recently been forcefully argued in the case of developmental biology by Claudio Stern (2022), and was humorously described in microbiology by Mark Sagoff:

> Bogglehead [a fictitious scientist] said the lunch crowd described sequence data as 'exhaust' since they are generated because they can be and not to test hypotheses. As the exhaust expands, it may spew a million correlations, but no [inkling] of causality. 'It's not the classic problem of searching for a needle in a haystack by adding more hay. It's that all the needles turn into hay,' Bogglehead said. (Sagoff 2019, p. 416)

As a point of departure, we'll see that *natural history*, that is, the classical observational branch of ecology, also views itself as menaced by these two threats. Both the importation of mathematical, abstract models, on the one hand, and the impact of the mass of readily available sequencing data, on the other, are regarded—by at least some authors—as casting doubt on natural history's present or future scientific *bona fides*.

In the case of the former, for instance, the steady advance of technical work in theoretical and mathematical approaches to ecology has been paralleled by the spread of what Sharon Kingsland has called a "generalizer" strategy across ecological study, natural history included. Her description of these two kinds of explanation is worth quoting at length, as we will return to a very similar idea below in Sect. 5.4:

> Historical explanations emphasize the way particular events accumulating over time have given rise to certain patterns, for instance patterns of species distribution and abundance. These explanations, often framed as narratives, tend to be cautious in offering predictions, focusing instead on how unusual events can change the course of history. Such events cannot be predictably generalized to other contexts. Generalizers, striving for predictability, may overlook the kinds of details that historical explanations focus on and concentrate more on processes that seem to operate everywhere. (Kingsland 1995, p. 218)

Skirmishes between authors who believe the historical and generalizing approaches to be incompatible date at least back to the mid-1960s, and have not abated in the decades since (Kingsland 1995, p. 202). Should ecology focus on patterns at the expense of natural-historical detail, or should it rather emphasize the importance of the contingent and the unpredictable to biological outcomes, possibly at the expense of being able to draw generally applicable conclusions?

There are a number of ways in which we might try to analyze this divide. Kingsland, for instance, notes that a complete historical account would have to invoke, among other things, the emergence of biodiversity as a central, organizing scientific concept over the 1990s as well as the desire for increased funding in ecology (Kingsland 1995, pp. 232–240).[1] Her survey was written before the "molecular turn" noted above, which has assuredly brought its own significant impacts.

While any consideration of these developments in the span of a single chapter is, therefore, doomed to be incomplete, I want to argue here that a central element in the concerns surrounding the fate and future of natural history involves disagreements in the scientific community about natural history's integrative promise. Does natural history give us the tools we need to respond to the biodiversity crisis? Is it a pivotal way to link bodies of scientific and practical knowledge, and to extend our understanding of the natural world across taxa, or does it rather lock us away in side-alleys, depriving us of the global picture needed for integration? Will the examples of real-world systems that natural history offers us actually be stringent enough to put those integrated theories to the test, or are we engaged in systematic cherry-picking? These are the kinds of questions that scientists have raised about natural history research, and they are precisely the kinds of questions that we have seen develop over the last four chapters.

To begin, however, we should start with just what natural history actually *is*, a question that we will see has no straightforward answer. I will then turn to the supposed "debate" over the status of natural history itself—both what it involves as well as skepticism that it is in fact taking place. Finally, I will explore the ways in which this worry about natural history invokes all of the central defining characteristics of integrative promise.

5.1 What Is Natural History?

It is common to locate Darwin himself not within a tradition of practicing "biologists" (which would itself be anachronistic) or even the new profession of "scientists," a term first coined by William Whewell in the 1830s (Ross 1991, ch. 1). Darwin is read rather as a "naturalist," a member of a largely amateur group of

[1] And invoking biodiversity draws our attention to the broader cultural context surrounding ecology, extinction, and crisis, an argument masterfully detailed by David Sepkoski (2020, ch. 6).

students of natural history, whose descriptions of living phenomena fill the early pages of journals like *Nature* (Baldwin 2015). These figures have been the subject of fairly extensive study in the history of biology, particularly thanks to Paul Farber (2000) and David Elliston Allen (1994, 1998). As Allen puts it, the kind of work done by naturalists had, since the seventeenth century, generally been the province of wealthy, motivated hobbyists:

> For almost two centuries there had been a few individuals lucky enough to be able to earn their livelihood from natural history, and the number of those had gradually been increasing. But on the whole those professionals differed from amateurs merely in the fact that they were paid to apply their expertise on a full-time basis: their outlook and their objectives were essentially the same as everyone else's. (Allen 1998, p. 361)

Allen's favorite example of a nineteenth-century figure who was indeed paid as a professional naturalist is George Bentham, the botanist whose gradual acceptance of evolution on the basis of Darwin's empirical work we saw in Chap. 3. His time at Kew Gardens turned him from an amateur into perhaps the leading professional botanical systematist of his day.

Despite the ups and downs of natural history study over the intervening several hundred years, we can still recognize many commonalities with the group that Allen describes. Natural history remains (with its closely allied cousin, taxonomy) one of the few places that amateurs can and do contribute significantly to professional scientific practice (Hopkins and Freckleton 2002). And the fundamental distinguishing feature of natural history remains careful, structured observation of the natural world. In that sense, it is still a sort of systematization of the same kinds of observational methods that any of us might use on a walk in the woods.

But from there, exactly what work gets to count as "natural history" comes in for debate. The broadest definitions take its "observational" aspect and extend this not only to characteristics and behaviors of organisms, but also to the abiotic environment and the paleontological record. Such is the definition offered by the pioneer of physiological ecology George Bartholomew:

> A student of natural history, or a naturalist, studies the world by observing plants and animals directly. Because organisms are functionally inseparable from the environment in which they live and because their structure and function cannot be adequately interpreted without knowing some of their evolutionary history, the study of natural history embraces the study of fossils as well as physiographic and other aspects of the physical environment. (Bartholomew 1986, p. 326)

The goal of natural history, then, is still to observe the characteristics of organisms—but to do so we must make reference to much more than organisms themselves. A similar, if much vaguer, impulse is found in a broad position paper on natural history (about which more later), which defines it as "the observation and description of the natural world, with the study of organisms and their linkages to the environment being central" (Tewksbury et al. 2014, p. 300). Environmental observation, once again, moves us beyond just the observation of organisms themselves.

Others expand our perspective to include the study of internal or physiological changes. Such is the view of the herpetologists Harry Greene and Jonathan Losos, for whom "natural history focuses on where organisms are and what they do in their environments, including interactions with other organisms. It encompasses changes in internal states insofar as they pertain to what organisms do" (Greene and Losos 1988, p. 458). This definition is at once narrower and broader than those we saw above: narrower in the sense that it does not encompass the abiotic environment or fossil evidence, but broader for involving physiology (a kind of observation, we might think, that is not simply a refinement of our everyday perception of the natural world).

Even more unclear definitions are often encountered across the literature. A recent winner of the ASN Distinguished Naturalist Award, the ecological geneticist Joseph Travis, defined natural history in his prize address as "the careful observation of nature, wherever nature is. Ultimately, it is what ecological, evolutionary, and behavioral science are supposed to explain" (Travis 2020, p. 1). For all that this definition replaces the vague term 'natural history' with the even vaguer 'nature', this idea of natural history as "raw material," as we will see, also sometimes lies behind fears over natural history's usefulness.

One way in which to see what these definitions have in common is to understand them as, at least in part, defining natural history contrastively with respect to its molecular and mathematical "competitors"—that is, as attempting to emphasize certain facets of the discipline that its practitioners believe have been lost in modeling or *in vitro* approaches. And here, it is time and again the crucial role of observation in natural contexts that comes to the fore. Bartholomew perhaps sums it up best: "At its most stereotyped, natural history has been, and is, strictly phenomenological. This is unexciting but not totally evil" (Bartholomew 1986, p. 326). In that sense, natural history sees itself as a part of an effort at accumulating, collating, and cataloging brute observations of the natural world, an effort that has already been in progress for centuries and, as was true for nineteenth-century naturalists as well, "appear[s] to be one of almost infinite duration, stretching into a very distant future" (Allen 1998, p. 362).

5.2 A Discipline Under Threat

If natural history is indeed part of a tradition of observation of the natural world dating back to Aristotle and beyond, why would its practice be the subject of any skeptical discussion at all? I'll consider below some authors who, indeed, doubt that any challenge to natural history even exists. But, at the very least, it is a remarkable feature of much recent literature on natural history that its practitioners *feel* as though they are under attack.[2] Rising emphasis on molecular methods, on

[2] Allen (1998) locates the beginning of this attack in the late nineteenth century, as the development of academic biology departments emphasized laboratory work and marginalized amateur contributions. I'll set aside this earlier shift, though it would be interesting to explore more thoroughly whether it parallels my story here.

mechanistic, *in vitro* studies, and on theoretical or mathematical modeling work all have been seen to pose varying kinds and levels of threat to what we might call "traditional" natural history.

In the late 1980s, Bartholomew already argued that increased interest in genetic and biochemical study in biology was crowding out space—and especially money—for natural history:

> In contemporary biology, much of the glamour and most of the funding go to research on the lower levels of integration. At these levels active researchers generally agree on the key questions. This consensus is so complete that we see large numbers of highly intelligent investigators with a treasure trove of instrumentation and techniques all concentrating on a few questions. This contrasts dramatically with research on the higher levels of integration, where we find many more good questions than investigators—not because of a paucity of investigators but, rather, a plethora of questions. (Bartholomew 1986, p. 328)

Before unpacking both the phenomenon that Bartholomew is describing and the explanation for it that he proposes, we can also see the same critique from Greene:

> The real impediments to natural history *sensu stricto* were then and still are more pervasive, sometimes subtle and sometimes not; they include excessive technophilia, a dearth of funding for discovery and descriptive research, the power of grant overhead monies in shaping academic programs, elitism on the part of some biologists, a shortage of journals that publish organismically focused studies and the widespread eclipse of biological diversity itself in high school and undergraduate curricula. (Greene 2005, p. 26)

The breadth of features that are implicated in these threats to natural history is quite impressive, and more than a little contradictory. Bartholomew argues that investigators with a wide array of technological methods focus their attention on a small number of questions, appearing to place the onus on the subjects of analysis rather than on the methodology. Greene, by contrast, accuses them of technophilia, which seems to locate the problem at the level of the methodology rather than the research question. Greene's invocation of a kind of disenchantment with the natural world itself, or a loss of interest in nature, is also echoed elsewhere (e.g., Tewksbury et al. 2014).

A sociological aspect is also underlined by both authors. Greene notes a trend toward "elitism," while Bartholomew argues that biologists often "believe that their own work is fundamental and that research in other disciplines, although sometimes technically ingenious, is either irrelevant or at best peripheral to understanding important biological problems" (Bartholomew 1986, p. 325). Of course, as Bartholomew's phrasing makes particularly clear, in some sense this is a tale as old as time: scientists very often do not understand the details of methods outside their own area, and lack of understanding breeds a lack of trust.

As for the important consequences, both seem to agree that, in short, resources are lacking. The best and brightest investigators do not work on natural history problems—at least, there are far more good questions than there are scientists capable of answering them. Neither do grant funding and journals support natural history scholarship to the extent that they should. Further, we see the threat immediately phrased by Bartholomew in terms of *levels of integration*. The lower we go down the organizational hierarchy, the story goes, the more resources are available.

So much for the empirical diagnosis of the case. While I have only demonstrated a few perspicuous examples here, when disputes arise concerning the importance, or lack thereof, accorded to natural history research, suffice it to say that this is the kind of thing that authors have in mind.

Of course, as I noted at the beginning of this section, one could interpret these trends differently, and under such alternative interpretations, natural history might not actually be under any threat after all. Stevan Arnold (2003), following Paul Farber (2000), argues that the general tendency has not been toward less study of natural history, but toward pursuit of natural history in more diverse contexts, as it gave birth to meaningful parts of zoology, botany, geology, behavioral ecology, molecular ecology, molecular evolution, bioinformatics, genomics, and more. All of these disciplines are what he calls "natural history derivatives" (Arnold 2003, p. 1066). In that sense, our problem is that we have attempted to fit contemporary biological practice into a much too constrained "natural history" concept:

> How can we reconcile these narrow, sometimes pejorative views of 'natural history' with a 250-year-old [naturalists'] tradition that includes the development of Darwinism and other major concepts as well as the incorporation of experimentation and model building? Why do we take a narrow view of 'natural history'? (Arnold 2003, p. 1066)

What natural history has done, according to Arnold, is not only to accumulate observations of the natural world, but rather to construct the concepts with which we analyze the natural world itself. From descent with modification to island biogeography to optimal foraging theory, a whole suite of concepts "are the triumphs of natural history, the essence of the discipline and its descendants," insofar as they are "conceptual tools for finding order in nature" (Arnold 2003, p. 1067). Natural history is defined by the pursuit of this goal, rather than any particular methodology. "The tools of the naturalist," he writes, "are equations and sequencers, as well as binoculars and notebook" (Arnold 2003, p. 1067).

In short, Arnold looks at the putative loss of prestige of natural history and draws the opposite conclusion: rather than the death of natural history, we instead have natural history *everywhere*, as it becomes virtually impossible for a contemporary biologist to do their job without being acquainted with a significant quantity of it.[3] Of course, he does so at the cost of radically enlarging the scope of natural history, to encompass any case in which we impose order upon complex and chaotic natural phenomena.

Greene takes an alternative approach to dissolving the issue. While I cited him just above in describing the problem, it's worth noting that, in fact, he doesn't believe that there is a genuine scientific conflict between natural history's supporters and its detractors. As he puts it,

> The dispute among us thus looks to be a red herring, an emotional but largely inconsequential misunderstanding that has perhaps been fueled in part by fuzzy, interchangeable use of the words 'theory,' 'models' and 'concept building' on the one hand and 'natural history,' 'organism-focused' and 'empiricism' on the other. (Greene 2005, p. 25)

[3] Thanks to Erin Saupe, who pointed out this response to a similar discussion surrounding the future of paleontology, which led me to appreciate its importance for natural history.

The very slipperiness in definition that we noted above, Greene argues, can be found across the supposed debate. Arguments that natural history proceeds without theory, or that construction of models is not organism-focused, require very particular definitions of all of these concepts—definitions that are not self-evident.

Finally, it's noteworthy that one can even be *empirically* skeptical about the presence of such a decline. In an analysis of eight decades of literature in ecology and conservation biology, Anderson et al. note that this literature has certainly demonstrated an increased rate of references to "many statistical paradigms (e.g., [the terms] *Bayesian*, *AIC*, and *MCMC*) and genetic analytical techniques (e.g., *phylogenetics* and *omics*) over the past three decades." But "this shift toward analytical techniques does not," they argue, "appear to come about at the expense of natural history—a loss that has been mourned and debated" (Anderson et al. 2021, p. 280). While this trend might, they argue, reflect a possibility for changes in ecological training (encouraging graduate students to develop proficiency in these technical tools, say), it does not, they think, indicate any attack on traditional natural history.

The state of the literature here is therefore somewhat perplexing. On the one hand, we have a pervasive sense among many practitioners that some kind of levels-of-organization prejudice has given undue weight to molecular and big-data biological work, while punishing organism-centric natural history.[4] On the other hand, we can be leery of this trend in at least two directions. In some sense, it was ever thus, with scientists arguing for the superiority and importance of their own areas of study.[5] And if we cannot clearly delineate the phenomenon that is cause for concern in the first place, it will remain difficult to evaluate it, or even to know to what extent it is genuinely occurring.

In what follows, I want to offer a novel but complementary diagnosis of the issue. As I noted at the beginning of the chapter, it will be my claim that this literature largely concerns the integrative promise (or the lack thereof) of traditional natural history. As such, worries about natural history's future are not ones over empirical data, but rather over values—to be more precise, they are worries about explanatory virtues. It is thus not surprising that they would only be expressed vaguely and with some difficulty in the scientific literature, which is not, as a rule, particularly well suited for conversations about these kinds of questions. I think we see this most clearly when we turn to the arguments that these scientists offer us for why these changes in the status of natural history should be viewed with concern, and how the problem might be remedied.

---

[4] This is all the more surprising given the fact that some authors, including Werner Callebaut (2012, p. 73), have argued that natural history is oldest of the "big data" disciplines; see Chap. 6.

[5] A point reinforced all the more by the fact that the argument of Claudio Stern in developmental biology reads exactly the same as that of the natural historians, with "mechanistic developmental biology" playing the role of natural history against the common villain of big data (Stern 2022).

5.3 Mounting a Defense

For the moment, then, let's follow these authors in assuming that there is indeed some kind of problem afoot in natural history. What do they argue makes natural history worth defending? As we will see, for a number of the most important players here, natural history is crucial because of its integrative promise. It offers us a site, more precisely, for vertical integration in the life sciences, one which (it is claimed) passes the tests of opportunism and stringency, and is uniquely suited to respond to the grand challenge of biodiversity loss.

We see this kind of account begin to take shape from the very first defenses of natural history. Bartholomew, who, as we saw above, phrases the threat in terms of levels of organization, describes the problem and its solution in terms of sites for fruitful integration:

> This problem is real, and its resolution, although difficult for many biologists to accept, is familiar. This resolution recognizes that there are a number of levels of biological integration, each with different properties and each offering special problems and unique insights. (Bartholomew 1986, p. 325)

At this point, however, it still isn't obvious whether he has in mind that biological knowledge inherently involves integration across levels, or if rather "integration" here means something more like "places where the biological action happens," each of which should be studied in isolation. This becomes clear a bit further on, where Bartholomew attempts to refine one of the definitions that we noted was somewhat slippery above: the idea that natural history is *organism-focused.* What is it that makes organisms so important to natural history? Of course, this is a question that is by no means alien to the philosophy of biology, the evaluation of and putative return to an "organismic perspective" having been the subject of much recent literature (Nicholson 2014; Walsh 2015; Shields 2017).

But precisely what makes organisms important for Bartholomew is that they function as a source of *pattern* in the natural world. To phrase the point in philosophical language, organisms are the site of the kind of repeatable causal interactions that enable us to draw general inferences. Just as Daniel Nicholson argues, in favor of a generalized organismic perspective, that "to understand any cellular or organismic phenomenon, it is necessary to situate the 'local' molecular process causally responsible for it within the 'global' context of the organized system that makes it possible in the first place" (Nicholson 2014, p. 353), Bartholomew argues that in the case of natural history, we "usually observe organisms, but linking these observations to populations, communities, and ecosystems can be based on such robust inferences that it is not possible to challenge them wholesale" (Bartholomew 1986, p. 326). It is the organism that makes vertical integration possible in the first place, and hence studies like natural history which privilege the organism are of inherent value. As he puts the point near the end of the paper,

> Knowing natural history allows an investigator to phrase questions with precision. It facilitates synthesis from lower to higher levels of integration and can help orient those biological sectors that focus on physiological mechanisms and issues far removed from the organisms they make up. (Bartholomew 1986, p. 328)

This tight connection between natural history's organismic focus and the possibility of integration is not peculiar to Bartholomew. Greene writes, as one of the four reasons that natural history is worth preserving, that

> Organisms themselves embody genetics, development, morphology, physiology and behavior, and they are the fundamental components of populations, communities and ecosystems. An understanding of organisms in nature is thus integral to studies at both lower and higher levels in the hierarchy of biological complexity.... (Greene 2005, p. 124)

And for Travis, one of two central ways to evaluate the "match of system to question" in biology is to ask "is the question relevant to the system?" That answer, in turn, "is a matter of natural history" (Travis 2020, pp. 4–5), whether we are exploring a question at the organismic, cellular, molecular, or populational levels.

To sum up, then, the first aspect of the defense of natural history offered by these authors points directly to vertical integration. Two features of this argument strike me as particularly notable, as they give us reason to believe that the authors see debates over natural history as ones involving its integrative promise. First, natural history research is presented as kind of condition for the possibility of vertical integration in the life sciences. We will not even be able to formulate reasonable biological questions, these authors contend, unless we can situate our understanding of those questions in the natural history of the individual organisms that are implicated. Thus, not only does natural history work itself have integrative promise, it is essential in order to legitimate the integrative promise of work at *other* levels of organization.

And second, it's worth stepping back and considering why this would be such a recurring trope in defenses of natural history. While the first science to feature in Ernest Rutherford's accusation of "stamp-collecting" was actually chemistry (Johnson 2007, p. 245), it is not difficult to imagine the same criticism being aimed at natural history. (I imagine that this is part of the impetus behind Bartholomew's quip that a phenomenological natural history is "not totally evil.") To declare that the accumulation of organism-centered natural history data is not only *not* stamp collecting, but is moreover crucial for our ability to vertically integrate in the life sciences is to stand this critique on its head.

Arguments in favor of natural history's stringency can be a bit harder to detect in the literature, but are nonetheless present. Another repeated mantra in these works concerns natural history's putative ability to *support* or *inform* hypothesis testing in the very fields that it hopes to integrate. If natural history is required as a contributing element to integrative explanations that bring together, for instance, ecological- and individual-level data, then natural history is instrumental in the construction of stringent tests of those very explanations.

This point is particularly emphasized by Travis (2020), who actually begins with the *limits* on the stringency of natural history. He starts with an example also dear to Kingsland (1995), MacArthur's broken-stick model for species abundance distributions. It is often observed that ecosystems tend to contain a small number of very common species, and a much larger number of relatively rare species. There are several distinct potential causal explanations for this. We might argue that this distribution is the result of natural selection: the most common species are (by far) the best adapted for the given environment, and rare species, while adapted well enough to exist in smaller numbers, are simply outcompeted. Or, alternatively, we might argue that a selectively neutral mathematical process could generate such a distribution, as advocated by the "unified neutral theory" of biodiversity (Hubbell 2001).

Taking yet a third option, MacArthur noticed that this kind of distribution could be produced on the assumption that an ecological niche was a finite resource that could be broken up into any number of different pieces, each representing the number of organisms in a species occupying that niche. Since a "breaking" process like this would lead to the distribution commonly seen in ecological observation, MacArthur proposed his own integrative hypothesis: namely, that the presence of the distribution should let us infer that abundance is governed by individual-level competition for a common resource that is being subdivided just like our stick. That hypothesis, however, failed. As we just saw above, a variety of different lower-level processes can generate this same higher-level distribution. More observation of species abundances by themselves, no matter how many times we confirm the presence of the ecological pattern, can simply never resolve the question of which process is at work in a given population (Travis 2020, pp. 2–3). Natural history cannot give us stringent tests in this kind of scenario.

But the case is different in two particular circumstances, where natural history is crucial for providing stringent tests of integrative theories. First, unlike the broken-stick distribution, in other situations, competing underlying processes might give rise to different, and incompatible, ecological observations, making natural history essential to legitimating our inference to the specific process at work. Second, natural history data might serve an essentially inductive role: if observations of hundreds of populations support a single hypothesis about the underlying process at work (despite, for instance, significant variation in the other influences on these populations and their environmental conditions), the inference to the integrated explanation from process to phenomenon might be directly legitimated (Travis 2020, p. 3).

Somewhat sheepishly, Travis notes that these phenomena are fairly rare, and moves on to ways in which natural history might not, in his terms, directly test a hypothesis, but rather merely "inform" or, even more weakly, "inspire" the pursuit of different kinds of integrative (and non-integrative) hypotheses in ecology, evolution, and so forth. But, in line with the integrative promise analysis that I'm offering here, I would propose a more positive interpretation of the situation. Of course, as we noted in Chap. 3, stringent tests will be rare, and thus the presence of a stringent test will be a significant feather in the cap of a putative integrative explanation. The fact that natural history does seem to serve this role in certain biological circumstances, however limited, forms an important part of the defense of natural history's integrative promise.

Opportunism is a bit easier to spot, thanks to the already mentioned complicated relationship between natural history and, in particular, molecular methods. As we saw above, Arnold (2003) attempted to bring all of this work under the umbrella of natural history itself. Further, Tewksbury et al. offer a wide-ranging defense of the utility and value of natural history, largely in terms of its contribution to our non-epistemic aims. They also offer four concrete propositions about what natural history practitioners should do in order to better position the discipline for the twenty-first century. The second such recommendation is to "avoid exclusivity, enhance connectivity, and embrace technology." As they elaborate, natural history's

> relevance will depend on the willingness of its practitioners to embrace new modes of observing the world and their capacity to recruit naturalists who use a much wider set of tools and skills than were historically associated with the study of natural history. The twenty-first century naturalist is as likely to work with a smartphone and a social network or a scanning electron microscope and a mass spectrometer as with binoculars and a hand lens. (Tewksbury et al. 2014, p. 306)

Part of this, of course, is about assuring the "relevance" of natural history in a competitive scientific landscape—they add, with the aim of convincing reticent undergraduates, that natural history is not only strolls in the woods and dead specimens in museum collections. But in light of our discussion, it is difficult not to read this as an appeal to opportunism. There are tools here, tools that will be used by potential collaborators whether or not they are contributing to the enterprise of natural history. Reconfiguring our science to take advantage of what they have to offer us is a clear opportunistic turn.

Finally, let's consider appeals to social and ethical values in defense of natural history. Like integration, these are widespread. Of the four reasons that Greene offers for defending natural history (the first of which we saw above), the third involves the fact that organisms "provoke widespread human curiosity," and the fourth is conservation (Greene 2005, p. 24). This preoccupation with value judgments dates back to Greene's earlier paper with Losos, where they argued first that natural history governs the worth that *biologists* place upon their objects of study. "It is because of phylogenetic systematics," they write, "that we can place special value on the coelacanth, the tuatara, and other 'living fossils,' and that we hypothesize that chimpanzees, not gorillas, are our closest relatives" (Greene and Losos 1988, p. 459).

But this process of valuing is broader than the scientific community, and extends to the general question of the place of biodiversity in human society. "For conservation to succeed," they argue, "wild organisms and places must have value in human society. The myriad interactions among plants and animals hold countless examples of novel solutions to problems with analogues in human welfare" (Greene and Losos 1988, p. 460). In his later work, Greene argues that it is this value dimension that remains as a point of contention between natural history's defenders and detractors, even after any potential epistemic or scientific disagreement is shown to be merely apparent. Taking aim at the paper by Arnold discussed above, Greene writes that

> Arnold expressed no concerns for the empirical and educational aspects of natural history *[sensu] stricto*.... Futuyma, Dayton and I [fellow advocates of natural history] are particularly concerned that we lack sufficient empirical reference points to move reliably among scales of time, space and biological organization, and that science therefore cannot adequately address environmental dilemmas. (Greene 2005, p. 25)

Here, then, not only are value judgments important to the promise of natural history, but the integrative case is connected directly to the value-laden advantages of natural history research: in order to address the biodiversity crisis, we need to develop integrative explanations, and to do that, we need to engage natural history. Equally explicitly, the manifesto of Tewksbury et al. is presented as a defense of natural history by describing its impact on four areas: human health, food security, conservation and management, and recreation. The summative section is entitled "Natural History's Place in Academia: Connecting Science and Society" (Tewksbury et al. 2014, p. 305).

Across these works, we see a variety of explicit arguments that the crises of biodiversity and conservation can only receive an adequate response if natural history is involved. This is precisely what we would expect if concerns about the future of natural history come down to questions about its integrative promise: its defenders are appealing, in part, to its ability to respond to one of the most pressing grand challenges currently invoked in the life sciences.

In contrast with the somewhat unexpected reference to integration, it is perhaps less surprising that natural history would defend itself by stressing its importance for securing global biodiversity. It is, after all, crucial to biodiversity preservation that we know and understand what it is that we are saving, and this is only possible via the kind of nuts-and-bolts observational work that natural history holds dear.

The idea that this is a genuine "grand challenge" is underscored by the also-repeated claim by these authors that natural history is a vital component of science education. Put differently, we need not only to adopt the insights of natural history in other parts of biological practice, but they need to be incorporated at the societal level as well in order to permit an effective response to the biodiversity crisis. This broader perspective places the emphasis on a discussion of biodiversity just where we would expect it to be if biodiversity is treated as a society-level grand challenge in addition to being a scientific question.

All of these value-laden elements are perhaps most pithily united in Greene's observation that "attitudes toward natural history—about what, if anything, needs fixing?—might hinge on the relative importance afforded to education and conservation as motivating goals for science" (Greene 2005, p. 26). What we think about natural history is not, in short, just a matter of how we evaluate its direct epistemic products. It is also a question of how we value the grand challenges which the knowledge that natural history generates might solve.

To sum up this section, then, there are a variety of characteristics of natural history that are taken by its supporters to make it worth defending. What makes those criteria interesting for our purposes, however, is that they slot very nicely into the structure developed thusfar for integrative promise. We see natural history lauded for its supposedly unique ability to enable inter-domain integration; for the power of natural history study to offer us stringent tests for the fields thus integrated, tempered by opportunistic collaborative efforts with neighboring research trends; and that work is taken to be crucial to global efforts to preserve biodiversity and respond to an urgent, socially relevant grand challenge. All that we might have hoped for to support an argument in favor of natural history's integrative promise.

In the rest of the chapter, I want to round out this argument with a kind of consilience of inductions, appealing to two further aspects of this discussion surrounding natural history that are likewise made much easier to understand if we take integrative promise to be central.

5.4 Ecology, Prediction, and Generality

While I lack the space and expertise to elaborate the case fully, I think that it is instructive to compare the consternation surrounding the nature and role of natural history with the discussion of generality and prediction in ecological models that has been explored by Alkistis Elliott-Graves (2018, 2019, 2022). In this field, obviously closely related to the pursuit of natural history, integration is also quite difficult to attain, and has been the subject of a heated argument—though here the focus is on horizontal, rather than vertical, integration.

As we saw above in Kingsland's historical account, and as has been extensively argued by Elliott-Graves, ecological work has for some time divided itself into two "camps." The first is a more theoretical branch, which "tend[s] to approach the complexity of ecological systems through highly abstract, mathematical, simple, and general models" (Elliott-Graves 2019, pp. 70–71). This corresponds to the "generalizer" approach that we saw Kingsland detail in the history of ecology. The other group, an applied branch, "tend[s] to deal with problems in particular systems, often using experiments or models that incorporate a lot of complexity" (Elliott-Graves 2019, p. 71). To be sure, these are extreme positions, and in reality they instantiate a spectrum, with most practitioners to be found somewhere between them. But the extreme positions are nonetheless often discussed by practicing ecologists themselves, who are embroiled in a debate over ecology's *predictive power*.

Ecology often produces surprising results. Studying the detailed natural history of a particular system regularly yields insights that general theory would not have led us to expect. Elliott-Graves offers us the example of "predator-prey role reversal," where a population of prey (in the first and now classic case study, whelks; Fig. 5.1) reaches such a density in an area that they become able to band together and attack their erstwhile predators (here, rock lobsters), often to the surprise or even horror of the observing ecologists; a further list of such unexpected results follows (Elliott-Graves 2019, pp. 66–67). The question with which ecologists often struggle is: what should we infer from this reaction?

Of course, some degree of surprise is to be expected in any science. But every feeling of surprise can be interpreted as a failed prediction from our prior (even if implicit) theory. As such, the more we accumulate surprising results, the more we might want to question the validity of the theories that are informing those tacit predictions. Unsurprisingly, one's position about the correct way to do this depends largely upon one's location on the theoretical-applied spectrum. Theorists argue that we have left too much data, too much evolutionary contingency, in our

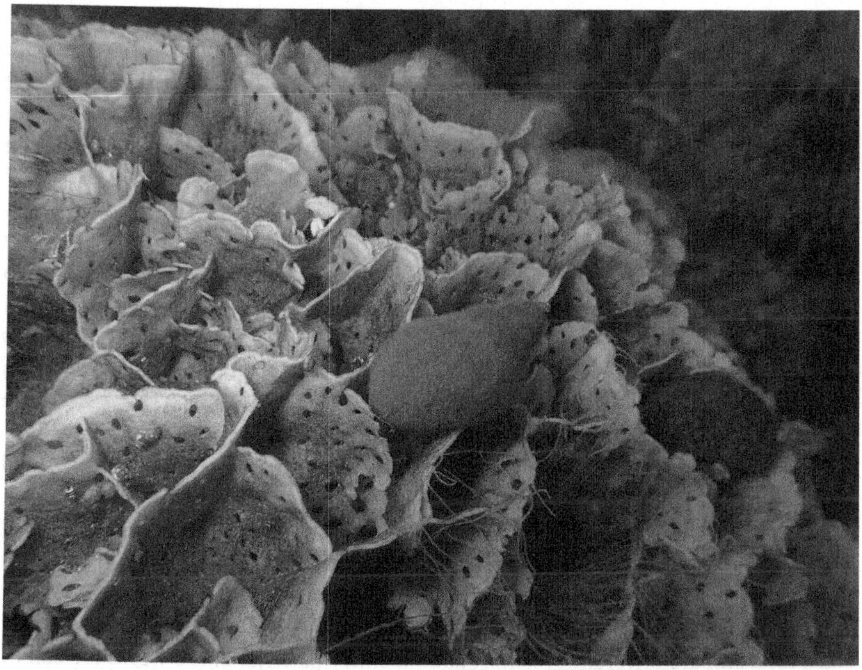

Fig. 5.1 *Burnupena papyracea* (center), a normally non-threatening species of whelk, in False Bay off the coast of South Africa. Wikimedia Commons, CC-BY-SA 4.0, by Peter Southwood

supposedly general ecological models, and need to remove these extraneous details to arrive at the causes that can be truly taken to pertain to any evolving population. In contrast, applied ecologists—often concerned first and foremost with the ability to practically intervene in ecology—argue that we should incorporate more detail into those models in order to enable effective intervention.

This much is Elliott-Graves's reconstruction of the state of play, which I find quite compelling. But I want to reinterpret it in light of our discussion thus far. The kind of integration for which theoretical ecologists are searching in this case is—in important contrast with our defenses of natural history in the last section—horizontal. They are looking for theories that can apply to radically different kinds of ecological systems composed of radically different kinds of organisms, experiencing different environmental pressures, access to resources, and so forth.

At the other end of the spectrum, while we don't find any of the authors of the defenses of natural history that we have canvassed in the examples of applied ecology elaborated by Elliott-Graves, the many instances of natural history work discussed by Tewksbury et al. very much come from this applied tradition—e.g., the consequences of the introduction of opossum shrimp into lakes in western North America (Tewksbury et al. 2014, p. 304).

I thus want to offer a tentative proposition for part of what is at issue in this debate. In addition to questions about prediction, or worries about the generality of ecological models, we also see in the conflict a disagreement about which *type of integration* is the most promising for further work in ecology. Should we attempt to add more detail, enabling the crucial role for vertical integration that authors like Greene have detailed above? Or should we remove that detail—likely cutting off our ability to vertically integrate—and focus instead on horizontal integration across different ecological systems?

Such a reading also has the virtue of easily encompassing Elliott-Graves's own proposed resolution of the debate. As she has argued, it is the *causal heterogeneity* of ecological systems that renders horizontal integration difficult. To express her position in the vocabulary that I've developed here, in other kinds of horizontally integrative enterprises, we are able to neglect unimportant differences in composition, causal interaction, and so forth, in order to construct theories with unifying power. In ecology, by contrast, "the differences between systems are much more often differences in terms of relevant causal factors, not mere details" (Elliott-Graves 2018, p. 1110). Ecology is thus an intrinsically difficult arena for horizontal integration.

Her prescription for a way forward is a kind of restraint: "A moderate level of generality, constrained to within particular types of phenomena or systems, is often sufficient to provide the explanatory and predictive power ecologists expect of their models" (Elliott-Graves 2018, pp. 1111–1112). Put differently elsewhere, scientists' goals to "compare different systems" and "identify patterns in the world ... can be achieved with quite modest generalizations" (Elliott-Graves 2019, p. 77). This fits quite nicely with the approach to integration developed in Chap. 2. Integration is not an all-or-nothing affair. Its extent can vary radically depending upon the integrative research program at issue, and, as already discussed, further integration is not an intrinsic goal to be pursued at all costs.

What's more, as she also notes, some approaches (in particular, "those used in the generation of systematic reviews and meta-analyses") "do not assume that generalizations hold, but *test* the scope of generalizations across different taxa/systems" (Elliott-Graves 2019, p. 78). Let's dig a bit into the details of the cases that she discusses in this context to see how they are related to the concept of stringency. Her central example is an article by Tina Heger and Jonathan Jeschke (2014), which describes the development of the *enemy release hypothesis*: the idea that invasive species will be more likely to be successful due to the absence of enemies in their novel habitat. Obviously, the hypothesis is extremely general, proposing an ecological mechanism that would hold across a wide array of natural populations.

What Heger and Jeschke propose in response to this extreme generality, then, is what they call the "hierarchy-of-hypotheses" approach. As they describe it:

> In a HoH, a broad formulation of a major hypothesis is positioned at the top level. This formulation needs to be broad and unspecific in order to capture all existing, more specific variants of the hypothesis. The broad hypothesis branches into more specific sub-hypotheses at the next level [and so forth]. On the lowest level of the hierarchy, each sub-sub-hypothesis is specific enough to allow rigorous empirical testing. (Heger and Jeschke 2014, pp. 741–742)

Again, there are a variety of ways in which to interpret this kind of methodology, but I think an extremely natural one is to view it as an actionable way to put putatively integrative explanations to stringent tests. In fact, when the authors tallied up the published tests of the more specific hypotheses found in the literature, the result was a genuinely mixed bag: about as much evidence to support the broad enemy release hypothesis as to reject it.

To sum up, the integrative-promise approach gives us an apt set of tools to analyze the distinction between theoretical and applied approaches in ecology. We see debates over which sorts of integration are in fact the most promising, as well as work attempting to find ways to put horizontally integrative hypotheses (which turn out to be more difficult to support in ecology) to stringent tests.

5.5 Who's Afraid of Natural History?

One last feature of the "debate" that we have discussed over the fate of natural history may already have become apparent to the careful reader. While we've seen a number of cases where the practitioners of natural history feel as though they are under threat, I have not cited any examples of authors arguing directly *against* the pursuit of natural history. Even Rutherford's famed "stamp collecting" quip was actually launched against the study of chemistry. Where, or who, is the opposition to the pursuit of natural history?

I want to explore here another advantage of thinking of this situation as one that concerns natural history's integrative promise. When we are considering an argument over competing explanatory virtues, I claim, we might well not see direct opposition to a body of research from those who nonetheless think that it lacks integrative promise. To be sure, sometimes scientists do take the time and effort to directly argue that their competitors' work lacks integrative promise. While I lack the space to reframe this case study in the present volume, I would now reinterpret a significant portion of the very public and vehement conflict over the utility of statistical methods in the study of evolution, about which I've written elsewhere and which I've mentioned above (Pence 2011, 2022), in terms of those methods' integrative promise.

But because integrative promise is both about the epistemic and the normative characteristics of a scientific research program, I think we have good reason to believe that an integrative promise debate will often look more like Kuhnian persuasion about the positive merits of a research program rather than a back-and-forth contest in the scientific literature. To borrow directly from Kuhn, debates over the integrative promise of potential explanations for phenomena within a science "involve the question: which problems is it more significant to have solved"—a straightforwardly normative question (Kuhn 1996, p. 110).

Nothing could be more clear from the citations we've seen from natural history's defenders above than that this is, indeed, one of the questions which animates our discussion of natural history's fate. In this sense, natural history's "detractors" aren't really detractors at all: they are rather arguing, by contrast, for the integrative promise of other kinds of enterprises to which we might instead direct time and energy. Rather than searching for arguments that natural history is unimportant or useless, we will find opposition in, for instance, the defenses of genomics that we saw at the beginning of Chap. 4, offering genetics as—just as we saw in the case of natural history—the key piece of knowledge enabling integration between biochemistry, physiology, health, and societal concerns. The appearance of a "one-sided" debate, then, is nothing out of the ordinary.

5.6 From Theory to Lens

This brings the first part of the book to a close. I began by introducing the general intuition behind integrative promise as a pattern of explanatory virtues that recurs throughout the life sciences. I then considered each of the elements of that pattern in turn: inter-domain integration, stringency, opportunism, and non-epistemic value judgments in support of grand challenges. Lastly, in this chapter, those strands came together for an analysis of the status of natural history.

This first part of the volume, then, has been intended to argue for the appeal of integrative promise as an analytic tool, at least in the examples that we've considered so far, including Darwin's works, the programmatic vision statement of the European Molecular Biology Laboratory, and the case of natural history. With luck, the reader is now at least convinced that in some situations, this pattern of explanatory virtues does indeed appear and is appealed to by practicing life scientists.

But this is only half of the story. Many approaches to explanation in the life sciences could be offered that are consistent with scientific practice, and so while perhaps of idle interest, this would not necessarily entail that integrative promise is a framework itself worthy of pursuit. To that end, in the second part of the book, I want to turn outwards, looking at a variety of broader instances that, as we got some hints of in this chapter, offer us explanations of the practice of the life sciences that illuminate features that we could otherwise have missed. Demonstrating appeals to the virtues that make up integrative promise across biological practice suffices, I claim, for legitimating that further investigation, but the real proof of promise's utility will lie in its ability to let us do novel philosophical work.

References

Allen DE (1994) The naturalist in Britain, 2nd edn. Princeton University Press, Princeton

Allen DE (1998) On parallel lines: natural history and biology from the late Victorian period. Arch Nat Hist 25(3):361–371. https://doi.org/10.3366/anh.1998.25.3.361

Anderson SC, Elsen PR, Hughes BB, Tonietto RK, Betz MC, Gill DA, Holgerson MA, Kuebbing SE, MacKenzie CM, Meek MH, Veríssimo D (2021) Trends in ecology and conservation over eight decades. Front Ecol Environ 19(5):274–282. https://doi.org/10.1002/fee.2320

Arnold SJ (2003) Too much natural history, or too little? Anim Behav 65(6):1065–1068. https://doi.org/10.1006/anbe.2003.2143

Baldwin M (2015) Making *Nature:* the history of a scientific journal. University of Chicago Press, Chicago

Bartholomew GA (1986) The role of natural history in contemporary biology. BioScience 36(5):324–329. https://doi.org/10.2307/1310237

Callebaut W (2012) Scientific perspectivism: a philosopher of science's response to the challenge of big data biology. Stud Hist Philos Biol Biomed Sci 43(1):69–80. https://doi.org/10.1016/j.shpsc.2011.10.007

Elliott-Graves A (2018) Generality and causal interdependence in ecology. Philos Sci 85(5):1102–1114. https://doi.org/10.1086/699698

Elliott-Graves A (2019) The future of predictive ecology. Philos Top 47(1):65–82

Elliott-Graves A (2022) What are general models about? Eur J Philos Sci 12:74

Farber PL (2000) Finding order in nature: the naturalist tradition from Linnaeus to E. O. Wilson. Johns Hopkins University Press, Baltimore

Greene HW (2005) Organisms in nature as a central focus for biology. Trends Ecol Evol 20(1):23–27. https://doi.org/10.1016/j.tree.2004.11.005

Greene HW, Losos JB (1988) Systematics, natural history, and conservation: field biologists must fight a public-image problem. BioScience 38(7):458–462. https://doi.org/10.2307/1310949

Heger T, Jeschke JM (2014) The enemy release hypothesis as a hierarchy of hypotheses. Oikos 123(6):741–750. https://doi.org/10.1111/j.1600-0706.2013.01263.x

Hopkins GW, Freckleton RP (2002) Declines in the numbers of amateur and professional taxonomists: implications for conservation. Anim Conserv Forum 5(3):245–249. https://doi.org/10.1017/S1367943002002299

Hubbell SP (2001) The unified neutral theory of biodiversity and biogeography. Princeton University Press, Princeton

Johnson K (2007) Natural history as stamp collecting: a brief history. Arch Nat Hist 34(2):244–258. https://doi.org/10.3366/anh.2007.34.2.244

Kingsland SE (1995) Modeling nature: episodes in the history of population ecology, 2nd edn. University of Chicago Press, Chicago

Kuhn TS (1996) The structure of scientific revolutions, 3rd edn. University of Chicago Press, Chicago and London

Nicholson DJ (2014) The return of the organism as a fundamental explanatory concept in biology. Philos Compass 9(5):347–359. https://doi.org/10.1111/phc3.12128

Pence CH (2011) "Describing our whole experience": the statistical philosophies of W. F. R. Weldon and Karl Pearson. Stud Hist Philos Biol Biomed Sci 42(4):475–485. https://doi.org/10.1016/j.shpsc.2011.07.011

Pence CH (2022) The rise of chance in evolutionary theory: a pompous parade of arithmetic. Academic Press, London

Ross S (1991) Nineteenth-century attitudes: men of science. Kluwer, Dordrecht

Sagoff M (2019) Can hypothesis-driven research survive the sequence-data deluge? Microb Biotechnol 12(3):414–420. https://doi.org/10.1111/1751-7915.13377

Sepkoski D (2020) Catastrophic thinking: extinction and the value of diversity from Darwin to the Anthropocene. University of Chicago Press, Chicago

Shields C (2017) What organisms once were and might yet be. Philos Theory Pract Biol 9(7):1–15. https://doi.org/10.3998/ptb.6959004.0009.007

Stern CD (2022) Reflections on the past, present and future of developmental biology. Dev Biol 488:30–34. https://doi.org/10.1016/j.ydbio.2022.05.001

Tewksbury JJ, Anderson JGT, Bakker JD, Billo TJ, Dunwiddie PW, Groom MJ, Hampton SE, Herman SG, Levey DJ, Machnicki NJ, del Rio CM, Power ME, Rowell K, Salomon AK,

Stacey L, Trombulak SC, Wheeler TA (2014) Natural history's place in science and society. BioScience 64(4):300–310. https://doi.org/10.1093/biosci/biu032
Travis J (2020) Where is natural history in ecological, evolutionary, and behavioral science? Am Nat 196(1):1–8. https://doi.org/10.1086/708765
Walsh DM (2015) Organisms, agency, and evolution. Cambridge University Press, Cambridge

Part II
Seeing the Life Sciences Through a New Lens

Chapter 6
Explanations Without Theories?

Big Data in Biology

Abstract While few biologists or philosophers have taken these pronouncements particularly seriously, some commentators have begun to see in the development of big data methods in the biological sciences the possibility that future biology will be able to do entirely without theory, simply drawing statistically robust generalizations from vast corpora of data and operating without understanding of underlying mechanisms or causation. To explore this claim in more detail, I'll unpack the idea of "data integration" that lies behind the construction of many of these big-data explanations, arguing first that data integration in this sense is not a species of the kind of inter-domain integration that I've developed here, and then that, while we might construct a plausible version of the argument in favor of theory-free biology, criticisms of that notion that have been advanced by philosophers clearly show us that it lacks integrative promise.

Keywords Big data · Machine learning · Theory-free biology · Databases · Data ontologies · Data integration

While it is dangerous to make proclamations about a century only twenty-five years in progress, it is nonetheless a fairly safe bet that the advent of big data and concomitant advances in statistical processing, machine learning, and similar methods will be one of the defining features of the twenty-first century. The sciences have, to be sure, not been immune to this wide, sweeping change. There are many reasons that the development of big-data-powered science might be important for a story of explanatory virtues in the life sciences, and I hope to touch upon several of them in what follows. But perhaps the most salient is a feature that is constantly cited by big data's defenders as one of its principal advantages: big data is often taken to, in some sense or another, be inherently about integration. Let's see some examples.

In the very first case study of her masterful volume on data-centric biology, Sabina Leonelli tells the story of iPlant (Goff et al. 2011). While its success ultimately proves to be a mixed bag, its very motivation is presented as integrative from the initial page of Leonelli's presentation:

C. H. Pence, *Integrative Promise*, Synthese Library 499,
https://doi.org/10.1007/978-3-031-88471-9_6

> iPlant is … a digital infrastructure developed in order to make various types of biological data travel far and wide, so that those data can be analyzed by several groups of scientists across the globe, integrated with yet more data, and ultimately help biologists to generate new knowledge. While recognizing that gathering all existing plant data under one roof is a hopelessly ambitious goal, iPlant aims to incorporate as many data types—ranging from genetic to morphological and ecological—about as many plant species as possible. (Leonelli 2016, p. 13)

The illustrated use-cases for iPlant talk about enabling scientists to "seamlessly integrate data across various formats, thousands of genomes, and hundreds of species" (Leonelli 2016, p. 14)—an appeal to horizontal integration—and "with data documenting other biological components, processes, and levels of organization"—adding vertical integration to the mix (Fig. 6.1).[1] She illustrates precisely the same kind of drive toward integration, as well, in the context of big-data research on *Arabidopsis*, the leading model organism in botanical research (for more on model organisms, see Chap. 7).

Similar points are raised by O'Malley et al., who argue that "the collection and analysis of vast amounts of data about DNA, RNA, proteins, and metabolites" (O'Malley et al. 2014, p. 813) has enabled integrated research into what they call "complex multilevel systems"; or Maureen O'Malley and Orkun Soyer, who write that "it is essential to develop multilevel interpretations and manipulations of datasets by addressing different levels of questions" in systems biology (O'Malley and Soyer 2012, p. 59); or Sara Green (2022), who notes that "network modeling and computational simulations in systems biology allow for the study of larger integrated systems and thus provide strategies for *recomposing* the findings in the context of larger systems." In a nutshell, one of the central purposes of big data is taken to be their usefulness for integration.

In this chapter, I want to explore this sense of integration and its relationship to integrative promise. The discussion will lead us to several different destinations. First, there is an inherent contradiction here, one that pits the challenges of integrating big data underlined by philosophers against a type of (supposedly) promising avoidance of theory often promoted by some of big data's most prominent advocates. In the end, I will argue that this dream of a putatively "theory-free biology" is impossible to realize. But the prospect, I think, should be taken seriously.

To do so, I'll first turn to a closer analysis of the idea of "data integration." Data integration has been taken by some authors (e.g., O'Malley and Soyer 2012) to be tightly connected with the kind of inter-domain integration that I introduced in Chap. 2. I will argue that this connection is merely apparent, the result of an

[1] While I am not going to develop a full case for the integrative promise of iPlant here, it's notable that Leonelli, drawing on the paper that introduces iPlant, emphasizes their argument that "understand[ing] how processes at the molecular level … affect, and are in turn affected by, the behavior, morphology, and environment of organisms" will "have the potential to foster scientific solutions to what Western governments call the 'grand challenges' of our time" (Leonelli 2016, p. 15). The iPlant introductory paper argues directly that "the iPlant [cyberinfrastructure] is a platform designed to enable researchers to make progress toward solutions of Grand Challenge problems" (Goff et al. 2011, p. 2).

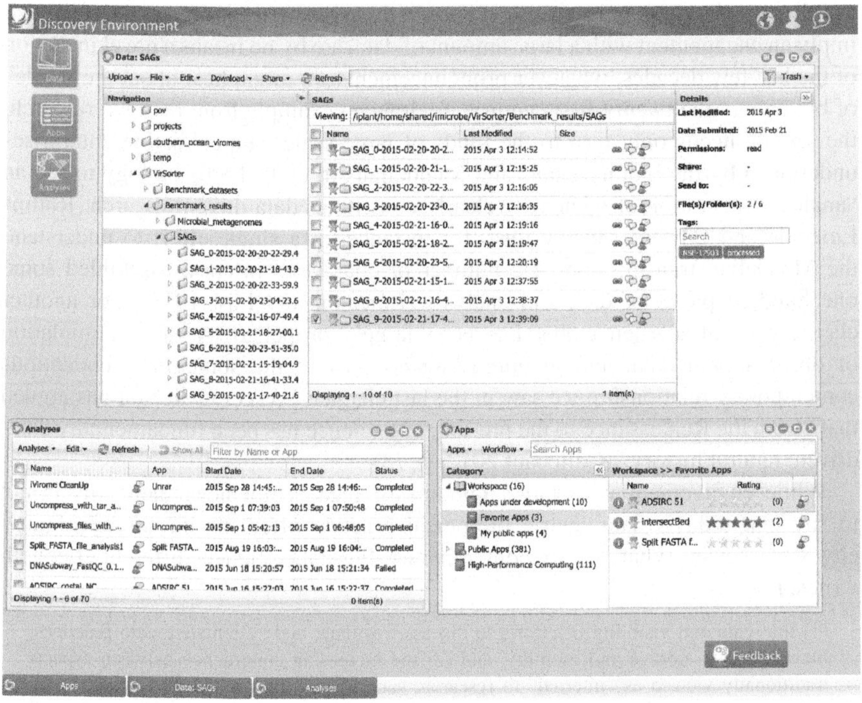

Fig. 6.1 Screenshot of the iPlant Discovery Interface. Figure 3 from Merchant et al. (2016), CC-BY 4.0

unfortunate polysemy of the term "integration" rather than a profound link between them. I'll then turn to a longer evaluation of the prospects for doing science "without" theories. Finally, I'll argue that the work of philosophers on questions surrounding big data in the life sciences demonstrates precisely that there is no integrative promise to be found in "theory-free" big data biology. But clarifying the reason that this is so will help us understand integrative promise itself by contrast.

6.1 An Apparent Contradiction

One of the most substantive contributions of analyses of the nature of "big data" biology, "data-driven" biology, or "data-centric" biology (all of which I will use as synonyms here) has been to underline that the "data" in "big data" are not a kind of neutral, readily available "natural resource" that simply exists to be harvested by scientific practitioners. Leonelli (2016), in particular, draws our attention to what she calls "data journeys": the ways in which those data must first be obtained, then cleaned, standardized, placed into databases, and—to put the point briefly—*turned into* the kind of thing that can be the object of scientific study. She notes, quite rightly, that the mere existence of scientific data, or the use of methodologies that imply an engagement with a large amount of data, are by no means a novel invention of the last few decades, whatever pronouncements we might hear about the "scale" of big data in the twenty-first century. To draw an example from my own research, the skeptic needs only turn to the kinds of large-scale data collection enterprises undertaken by the biometricians and Mendelians in early 1900s biology to see an "analog" illustration of what very clearly constitutes data-driven research. Rollins Emerson and Edward Murray East, for instance, in a single effort to understand the Mendelian transmission of a number of characters in corn, assembled some one hundred pages of data and analysis (Emerson and East 1913). For another classic case of a science that has been largely propelled by the accumulation of observational data, and in interesting contrast to the sometimes tendentious contemporary relationship we saw in the last chapter, Werner Callebaut has argued that "from the Renaissance to the present, one science has been fundamentally data-driven: natural history" (Callebaut 2012, p. 73).

While engagement with large amounts of data is not new, then, this is not to say that today's data-centric biological practice offers us no source of novelty whatsoever. What makes contemporary big-data science different, Leonelli writes, is

> (1) the scale and visibility of efforts to discuss, regulate, and systematize data practices, including data storage and mobility, and (2) the increasing congruence between aspects traditionally viewed as "internal" to research, such as the ways in which organisms are conceptualized and the procedures adopted to investigate them, and "external" aspects such as the embedding of research practices into specific social, cultural, and economic structures and trends. (Leonelli 2016, p. 173)

In short, what makes today's big-data biology distinctive is the thoroughgoing recognition that data are not produced only for the pursuit of narrow epistemic goals in the context of one single laboratory. While Emerson and East might have hoped that their data could prove of use for other agriculturalists and biologists interested in heredity, they did not package their work for immediate public consumption—they did not have an "audience" in mind for their data, with a concomitant set of standardized data formats, an online database into which they planned to deposit their work, and avenues to discuss and debate with their colleagues not their results, but the very nature of the enterprise of gathering and sharing data.[2]

Focusing on these data journeys allows us to realize, among other things, just how contingent many of these choices in fact are. Among other reasons, these processes are "embedded within resilient political, cultural, and economic structures," which entails that their epistemic, political, and ethical valence turns "on how data journeys are managed and financed and who is involved in their development" (Leonelli

[2] While considering this story in detail would take me too far afield here, precisely this absence was lamented particularly by Karl Pearson, who feared that lack of standardization in vocabulary and methods of presentation would render interpretation of statistical data nearly impossible (Pearson 1894; Weldon et al. 1901).

2016, p. 164). No data can be rendered suitable for every single imaginable public, and so the details of the data journeys at issue—"the choices that [scientists] make when considering which data to view as most relevant to their research, how to integrate those data, and which expertise to involve in that process" (Leonelli 2016, p. 160)—will have immediate and concrete effects on the character of that knowledge.

Let's step back and consider this through the lens of integrative promise. Leonelli's worries about the ways in which data journeys impact the epistemic quality of big-data biology underline a point that I noted frequently in Chap. 2: inter-domain integration is an extremely complex and hard-won empirical achievement. It will depend, here, upon choices made by individual scientists working at the bench (e.g., should we gather data on a measure that, while not of immediate use to us in the lab today, could prove vital for future integrative efforts?) as well as by collective groups of scientists working to establish common characteristics for the kinds of data that will be considered as acceptable in large-scale databases (e.g., do we target this database only at our own community, or do we make it interoperable with the datasets of other groups, which might spur unexpected inter-domain integrations?).

Further, we see clear indications of the tradeoffs between stringency and opportunism that I discussed in Chap. 3. What makes data useful to a particular scientific community involves not just the data themselves, but also reflection on precisely the choices that have already been made within the data journeys that have led to the present moment. Put differently, biologists in data-centric fields spend a significant amount of time trying to balance the opportunistic reuse of already-gathered data—after all, the data are available right here in my office, without my even having to perform any experiments—with their putative stringency for the research purpose to which they would like to apply them. Will the data actually be able to test the quality of the integration that I've targeted, or will I have taken on too many tacit assumptions that render moot my effort to integrate? This, too, is a widely recognized feature of data-centric biology. O'Malley et al. have argued, to which I will return below, that "the success of these approaches depends on a judicious choice of physical entities and interactions within the systems of interest and their incorporation into models that abstract away from details or idealize features" (O'Malley et al. 2014, p. 820).

Lastly, Leonelli's frequent references to the embedding of these results in their social and political contexts indicates their importance to the kind of grand-challenge resolution that I described in Chap. 4. As she puts it, in dedicating time to the pursuit of these socially and ethically important outcomes, scientists "are investing a significant amount of their resources on producing results that ... are primarily meant to serve a wider social agenda" (Leonelli 2016, p. 158). On a first impression, at any rate, it seems as though the standard approach to data-centric biology in recent philosophy of science dovetails nicely with its being defended on the basis of its integrative promise.

6.1.1 Theory and Data in the Life Sciences

But my discussion here has been sloppy in two ways that it is important to clarify, one of which has been the subject of much controversy in contemporary literature on data-driven science. (Both of these issues were also the topic of some discussion in Chap. 2.) First, what exactly is it that is being integrated in these contexts? In Chap. 2, I argued that integration should be interpreted very generally as connecting *domains*, permitting my approach here to encompass both horizontal integration across the tree of life and vertical integration across different levels of organization. But very often, integration in the big-data context is taken to directly concern instead *data integration*, a more limited process that involves interoperability between different bodies of big data. What is the relationship between data integration and inter-domain integration? I will return to this question in the following section.

In this section, however, I want to take a slightly different approach: of what exactly is integration taken to be a property? I pose the question because it reveals, I think, an apparent contradiction between (1) my emphasis on stringency as an explanatory virtue (and an element of integrative promise), (2) the perspective on big data developed by Leonelli, and (3) one of the advantages often taken to be entailed by big-data research. In slogan form, we might phrase the question as: does big-data biology let us do biology *without theory*?

In Chap. 1, when I first introduced integrative promise, I described it as a collection of *explanatory virtues*. That is, I take inter-domain integration (as well as stringency, opportunism, and appeals to non-epistemic values) to be a property of explanations, rather than of theories. In part, as was mentioned in passing, this was to avoid begging the question concerning exactly this controversy. Leonelli takes this kind of integration to be a property of scientific theories (Leonelli 2016, p. 122), in the sense of Mary Hesse's understanding of theories as networks of interrelated terms (Hesse 1980). Those advocates of data-driven biology who claim that it can be practiced without any theoretical commitments at all certainly do not ascribe integration to theories—although, in part because this claim is essentially never defended in the philosophy of science, exactly what they take integration to involve is much more complicated to spell out, as we'll see in what follows.

It is rare that a short editorial in a popular technology magazine leaves a lasting imprint on the philosophy of science. In the July 16, 2008 issue of *Wired* magazine, editor Chris Anderson, whose prior editorial credits include time at both *Science* and *Nature*, penned a brief article entitled "The End of Theory: The Data Deluge Makes the Scientific Method Obsolete" (Anderson 2008). It is difficult to imagine a more straightforward fashion in which to provoke philosophers of science, from the invocation of "The Scientific Method" as a single, proper noun to Anderson's claim that "companies like Google … don't have to settle for models at all" of the phenomena which they study.

The idea is simple enough (although at least one commentator describes it as "so obviously flawed that I wouldn't have referred to it at all hadn't it become so influential"; Callebaut 2012, p. 74). With a large enough dataset, Anderson argues, we no longer have to search for causal links in our data. We can simply generate predictions from our corpora of data, cutting out the *tertium quid* of scientific hypothesizing or theorizing.

> This is a world where massive amounts of data and applied mathematics replace every other tool that might be brought to bear. Out with every theory of human behavior, from linguistics to sociology. Forget taxonomy, ontology, and psychology. Who knows why people do what they do? The point is they do it, and we can track and measure it with unprecedented fidelity. With enough data, the numbers speak for themselves. (Anderson 2008)

As a concrete example, Anderson mentions Craig Venter's early work with high-throughput DNA sequencing. Sequencing vast tracts of open ocean allowed Venter and colleagues to, in their words, identify "148 previously unknown bacterial phylotypes" (Venter et al. 2004, p. 66). Should you say to yourself, Anderson writes, that this doesn't amount to a genuine discovery of new species, in our complete ignorance of the ecology, behavior, distribution, or phenotypes of these organisms, then "you may be stuck in the old way of doing science." Identifying these bacterial species "with Google-quality computing resources" means "Venter has advanced biology more than anyone else of his generation" (Anderson 2008).

It would be tempting to write off Anderson as someone who had imbibed a bit too much Silicon-Valley atmosphere, but claims not too dissimilar from his are found in more staid and reflective corners of the scientific community. Gert Jan van Ommen, at the time head of the Center for Human and Clinical Genetics at Leiden, wrote in 2008 that

> if it becomes cheaper to just collect *all* data required than to run after a hundred consecutive, plausible, but wrong hypotheses, starting with a hypothesis becomes an economic futility. The hypothesis as a guiding principle is then replaced by a truism: if one does not throw away anything before thoroughly assessing its irrelevance, one will always find what one is looking for. (van Ommen 2008, p. 1)

Moreover, as we saw in the last chapter, big-data biology is at least occasionally set up as a kind of villain against which more "traditional" biological practice can be defined and defended. In that sense, even if "the end of theory" is a product of Silicon Valley hype, an end-of-theory style argument is nonetheless taken either to be actually or potentially problematic for the future of biological science. George Bartholomew, whose defense of natural history was extensively considered in Chap. 5, also argued that

> physiologists studying hibernation have all too often been indifferent to its biological complexity and context in natural history. They have conceived it primarily in terms of mammalian adjustment to seasonal cold, and their understandable fascination with the mechanisms of mammalian temperature regulation has led to an overwhelming accumulation of data, data whose sheer volume has deflected our thoughts away from the central biological issue. (Bartholomew 1986, p. 327)

Notably, and providing more evidence for my claim that the epistemic challenges of big data are not entirely novel, the same kinds of worries can be easily found in early twentieth-century philosophy of biology (Nicholson and Gawne 2015; Sarkar 2023). In 1929, in his *Biological Principles*, Joseph Henry Woodger wrote that "we are, therefore, in danger of being overwhelmed by our data and of being unable to deal with the simpler problems first and understand their connexion. The continual heaping up of data is worse than useless if interpretation does not keep pace with it" (Woodger 1929, p. 318). Three years later, Joseph Needham argued that "it is extremely important to realize that we need theoretical biology at least as much today as the more practical studies of experimental workers, and until we get it, the structure of biology will remain, as has recently been remarked [by Woodger and Needham himself], a medley of unorganized *ad hoc* hypotheses" (Needham 1932, p. 84). Ludwig von Bertalanffy, a year later, argued that "when we ... glance at the thousands of experimental investigations which are published yearly, we cannot avoid the heretical opinion that it is perhaps not so necessary to add another dozen or so to these as seriously to set about the task of theoretically exploiting the mountain of raw material we already possess" (von Bertalanffy 1933, p. 26).[3]

In the contemporary literature, given the significant funding and hype surrounding big-data research, the tone has become even more moralizing. Writing in the case of developmental biology, Claudio Stern laments the rise of a data-driven biological culture, arguing that

> The casualty is not just experimental design, but the whole Waddingtonian ethos of uncovering *rules* that govern the behavior of cells, tissues and systems. In the last 20 years the community has become obsessed with massive data collecting and semi-automated (computational and statistical) methods of "self-organization" of the data generated, but with little integration to attempt to understand mechanisms. (Stern 2022, p. 31)

As was the case with the last chapter's argument surrounding natural history, setting aside cases like Anderson (and similar figures, often trained in the physical sciences, like Stephen Wolfram; e.g., Wolfram 2021), it is hard to find a figurehead in contemporary biology or philosophy of biology who defends the obsolescence of theory. But the idea that big data is nevertheless threatening to the progress of the contemporary life sciences is clearly keenly felt and widely shared. In that sense, even the mere possibility of theory-free biology is something that we owe it to ourselves to take seriously.

From our perspective, then, can we reconstruct a view of science that would supposedly be in play for someone arguing in favor of theory-free biology and, in turn, understand where exactly integration should be ascribed? Of course, on pain of a self-evident contradiction, integration cannot be borne by theories. Unpacking the authors that we've canvassed so far (whether supporting a theory-free view or inveighing against its menace), we see a number of different features of science picked out as especially important. Let's consider whether or not each one helps us understand the nature of integration in data-driven biology.

Hypotheses van Ommen (2008) and Anderson (2008) both disparage the usefulness of *hypotheses* in scientific practice. Perhaps most (or too) charitably, we could read this as an argument against particular kinds of methodological choices in the construction of statistical experiments—moving from a regime of hypothesis-testing to discovery in statistical models. As this move is purely negative, however,

---

[3] Thanks to Dan Nicholson for bringing these authors to my attention.

it seems as though it doesn't give us an answer to our question: stating that it is *not* hypotheses doing the integrating, or that we ought *not* use hypothesis-driven inferential methods, doesn't tell us what we ought to do instead, or in what sense data-driven science is integrative. Moreover, any generalized argument against hypothesis-driven science must reckon with a now fairly extensive body of literature claiming that it is precisely a return to hypothesis testing that can, in part, help scientists resolve the replication crisis (Nosek et al. 2018; though see also Ledgerwood 2018).

We might be able to get a little more clarity, however, by revisiting Needham. Recall that Needham warned that, without theory to bind it, biological reasoning might resemble "a medley of unorganized *ad hoc* hypotheses" (Needham 1932, p. 84). That is, what matters is not whether or not hypotheses are proposed—Needham implies here, quite rightly in my opinion, that even the most data-driven science will nonetheless produce generalizations concerning the data thus far observed that, when extended to unobserved cases, will have something of the character of hypotheses. The question is whether or not these hypotheses are *ad hoc*. This, in turn, is underspecified, and could be further explicated in a number of different ways.[4]

While spelling out the full details of these different theories of *ad hoc*-ness would take us too far afield for my concerns here (and at the price of being a bit simplistic in presenting the alternatives), one thread can be seen to unite a number of different accounts. In evaluating whether or not a hypothesis is *ad hoc*, we often look beyond the confines of the hypothesis itself to the role that it plays in the explanation of the phenomena that it putatively explains. Consider, for instance, Karl Popper's classic presentation of the idea, on which *ad hoc* hypotheses are "almost circular," insofar as they do not "have a variety of testable consequences, and among them, especially, testable consequences which are different from the" phenomena that we would like to explain (Popper 1983, p. 133). Put differently, such a hypothesis "goes as little as possible beyond the facts it is expected to explain" (Popper 1983, p. 231).

Or consider an alternative view of *ad hoc*-ness, invoked among others by Jarett Leplin (see Schindler 2018, pp. 126–127), on which, to borrow Schindler's formulation, "ad hoc charges have got something to do with lacking unifiedness" (Schindler 2018, p. 127). (Leplin himself makes this a part of his claim that *ad hoc* hypotheses are evidence of a theory's being "non-fundamental.") Unification, as we have already seen and will see again just below, is of course commonly evaluated in the context of explanations. Schindler's own theory, which he calls the "Coherentist Concept of Ad Hocness," has it that a hypothesis is *ad hoc* if neither our theory nor our background knowledge "provides *good reason* to believe" it (Schindler 2018, p. 133). Yet again, there is a natural explanatory reading of this "good reason" cause.

[4] Many of which have been pursued in the literature; see Schindler (2018, pp. 120–141) for a helpful survey.

To stay with this common thread for a moment, then, we could therefore read Needham's claim as invoking the explanatory credentials of these hypotheses. If this reading of Needham is right, then, we should turn from the role and nature of hypotheses to the kinds of explanations that might be at work in theory-free biology. What would make the proposition of theory-free hypotheses problematic, for Needham, would be their reliance on these broader explanations that this data-driven science is taken to offer us.

Explanation: Mechanism and Causation Before we explore concepts of explanation that might be at work in theory-free biology, however, we should set aside one common notion of explanation as *not* that which the theory-free biologist intends to discover. Almost as often as we encounter distrust of theories and hypotheses, we find distrust of causality and mechanism. Anderson, for example, writes that "correlation supersedes causation, and science can advance even without … really any mechanistic explanation at all" (Anderson 2008). Despite the emphasis placed upon mechanistic explanation by a number of contemporary philosophers of biology (Machamer et al. 2000; Glennan 2002; Craver and Darden 2013), for anti-theory authors, unearthing reliable correlation is sufficient to guarantee the *bona fides* of a scientific explanation.

It will not take extensive argument to convince the reader that this is a significant departure from the literature on scientific explanation which, whether one is monist or pluralist about the notion, and whether one countenances non-causal explanations in addition to causal ones (Baker 2005; Skow 2014), has generally followed Wesley Salmon in arguing that, at least for a very important class of scientific explanations, "we explain events by showing how they fit into the causal nexus" (Salmon 1984, p. 276). Renouncing such a commitment will not come easily. That said, it is not unprecedented. Let's explore a few ways in which we might accept a genuine, explanatory role for correlations drawn from big data in the absence of mechanism or causation.

Explanation: Unification The first is among the classic theories of scientific explanation: generalizations, even if merely correlative, might be explanatory insofar as they give us a way in which to unify a wide array of disparate observations of phenomena, a theory of explanation now often ascribed to Philip Kitcher (1981). Even if the generalizations do not grab onto "real causes" in the world, we might nonetheless think that such correlations could allow us, in Kitcher's words, "to determine the conditions which must be met if science is to be used in answering an explanation-seeking question *Q*" (Kitcher 1981, p. 510). Those answers, he argues on the basis of historical examples from Newtonian mechanics and Darwin's theory of evolution by natural selection, are such that they use a small set of "similar arguments in the derivation of many accepted sentences" (Kitcher 1981, p. 519). Intuitively, Newtonian mechanics explains because we see that a whole host of phenomena in the world are the result of initial conditions plus forces; Darwinian evolution explains because a plethora of features of the biological world are the result of natural selection, genetic drift, and other evolutionary factors acting on variation in populations over time.[5]

It's clear that it's at least conceivable that explanations that fail to invoke causes could still unify in Kitcher's sense. To take a particularly obvious example, phlogistic chemistry seems to have successfully unified phenomena in exactly the way that Kitcher describes, while—at least as we think today—not having accurately identified any real, causal agent at work in the world (although Kitcher gestures at ways in which we could try to consider this a "spurious" unification; Kitcher 1981, p. 529). In this sense, theory-free biology might be showing us ways in which "patterns" emerge from scientific data that can serve to unify those data without offering us a causal picture of how the world operates. Those patterns would gain in explanatory power when they make it possible to more effectively unify a large number of phenomena, and would become *ad hoc*, on the contrary, if they only "unified" little beyond the body of data that prompted their proposal. This seems to be a potential candidate for "theory-free" biological explanations. (I will return to the evaluation of these candidates further on; for now my aim is simply to effectively play devil's advocate.)

Prediction and Pragmatism A more radical response, but nonetheless in line with the arguments that we find for theory-free biology, would be to abandon the pursuit of explanation entirely. (Perhaps it, too, is tied to Anderson's "old way of doing science.") In the philosophy of science, we have one partial precedent for this kind of view. While acknowledging that explanation is still an important goal for the philosophy of science, Bas van Fraassen's anti-realist, pragmatic understanding of scientific explanation makes the pursuit of explanations secondary, and parasitic on the basic pursuit of empirical adequacy.

Van Fraassen starts with a very similar approach to Kitcher regarding the central question that a theory of scientific explanation needs to ask. Explanation consists in a demand by someone to know why a particular phenomenon occurred, rather than another member of a possible contrast class: "Why P, in contrast to X?" More precisely, a demand for explanation is the posing of such a question in a context where "(a) P is a member of X, (b) P is true and the majority of X are not" (van Fraassen 1977, p. 149). Further, in order to avoid a number of problems of explanatory asymmetry (*without* necessarily making reference to causes, which van Fraassen wants to abandon as a result of the apparent breakdown of causation in some areas of quantum mechanics), we also need to propose a relation of *explanatory relevance*, which determines which facts will be held constant for the purposes of evaluating a given demand for explanation.[6]

[5] I will not enter into the fine-grained details of Kitcher's picture here, not least because he introduces a property of explanations, also called stringency, but used in a radically different sense than my use of the term.

[6] More precisely, van Fraassen fleshes out this relation in terms of the distance between possible worlds, though this is not important for my purposes here.

But importantly, on this view, *scientific* explanation is in fact only distinguished from other kinds of everyday explanation by the peculiar characteristics of the relationship of explanatory relevance which science proposes—in van Fraassen's words, "scientific theory enters mainly in the evaluation of possibilities and probabilities, which is only part of the process, and which it has in common with other applications such as prediction and control" (van Fraassen 1977, p. 150). More radically, he writes, since scientific explanations are simply one species of everyday, pragmatically useful demands for explanation, strictly speaking "there are no explanations in science" (van Fraassen 1977, p. 150).

Such a view harmonizes quite well with the pursuit of theory-free biology. If the goal of science is not to uncover causal truths, but rather to alight upon regularities that enable prediction and control of "medium-sized dry goods" (to use J. L. Austin's much-copied phrase), then there is nothing in the way of our arguing that a particular demand for the explanation of a given phenomenon (e.g., the distribution of bacterial species in the Sargasso Sea) might be satisfied by an explanation in terms of a correlation pattern found within some large dataset (e.g., the output of Venter's high-throughput DNA sequencer). To be sure, the precise details of the explanatory relevance relationship that makes this an acceptable response in context would need to be worked out in greater detail, but this is no larger a problem for a putative correlative big-data explanation than it would be for an explanation in, say, quantum mechanics or organic chemistry. We have here, then, another viable candidate for the nature of theory-free biological explanations.

In sum, then, it remains true that the most promising locus for an evaluation of integrative promise in the case of "theory-free biology" would be as an explanatory virtue—a property of either Kitcher-style unifying explanations or of van-Fraassen-style pragmatic explanations. While we will find neither causal-mechanical explanations nor theories in such a biology, we could plausibly find explanations in these two broader senses, and these could thus have—or fail to have—integrative promise. Such a claim has the advantage of being consistent with the limited defenses that exist of theory-free biology, as well as being directly attacked by the critiques of theory-free biology from philosophers like Needham and Leonelli. When we move to critiques of theory-free biology, we will thus be doing so from what I take to be the most charitable reconstruction of the position available. Evaluating the overall case for the integrative promise of theory-free biology will be my project in Sect. 6.3. Our next target, however, is to examine those counterarguments in greater detail.

6.1.2 Resisting Theory-Free Biology

The absence of any extensive philosophical or biological literature defending a theory-free view of data-centric biology is already indication enough of the relatively low regard in which such a view is held by most philosophers of biology. What arguments do they offer?

Callebaut sets out what is likely the most straightforward rejoinder to Anderson: the claim that knowledge in central, data-driven areas of the life sciences like epigenetics and evolutionary developmental biology was "acquired without resort to modeling … is nonsense" (Callebaut 2012, p. 75). It is certainly the case that models become more intricate in these domains. He borrows the term "*multiscale modeling* … [i.e.,] any kind of modeling, computational or other, that includes (1) multiple length scales, viz., components from two or more levels of biological organization … and/or (2) multiple time scales" (Callebaut 2012, p. 78) to describe the kinds of particularly complex models that result from integrative enterprises engaging with big data. Understanding how multiscale modeling can actually succeed—or, put differently, how "we can have science as an engineering practice, *and* it can help us to understand the world" (Callebaut 2012, p. 79)—is a non-trivial philosophical enterprise. But succeed it indeed has, and it has done so by the construction of precisely the kinds of models and theories that the advocate of theory-free biology tells us we do not need. Actuality implies possibility.

But the real-world existence of big-data biology *with* theories and models implies that we need to engage with data-driven biology not in the abstract, but in the form of concrete case studies. I lack the space to develop one in detail here, but we can draw some morals from a few other authors. First, I have already quoted from O'Malley et al., who mention in evaluating integrative approaches that achieving them "depends on a judicious choice of physical entities and interactions within the systems of interest and their incorporation into models that abstract away from details or idealize features" (O'Malley et al. 2014, p. 820). A clearer apparent invocation of stringency in my sense would be difficult to find.

Second, Leonelli has unpacked the relationship between data as found in a biological database and both scientific experiment and scientific theory. She begins by offering a compelling argument that data curation—labelling, processing, and packaging data for travel between contexts—is a form of "*theory making*; and … its products, such as bio-ontologies, developmental stages, or taxonomic systems, constitute *classificatory theories* that can be clearly distinguished from other forms of theorizing typically discussed within the philosophy of science" (Leonelli 2016, p. 115). For Leonelli, theory-free, data-centric biology is thus a kind of intrinsic contradiction: claiming that such a thing exists renders invisible the difficult work that is usually required to make data function in the scientific process.

Moreover, and thirdly, she argues from a number of examples that the ability to interpret the value of the data found in a database as evidence for a given scientific claim requires a kind of tacit background knowledge (*sensu* Polanyi) "that cannot be acquired without personal involvement in, and extended familiarity with, certain types of material intervention. Learning happens as much through reasoning as it does through doing, and experimental intervention remains the most important form of doing within the biological sciences" (Leonelli 2015, p. 106). Collating a vast number of data points, then, is not the *only* thing required, as efforts to ascribe further evidential value to data are often not the result of one integrative enterprise, but rather of fragmented efforts by a variety of research teams (Leonelli 2016, p. 109). Understanding whether or not the kinds of tacit knowledge encoded within a database are compatible with a given effort at integration requires precisely the kind of familiarity with the nuts-and-bolts details of the biological material that Anderson claims we can dispense with.

On an initial reading, then, the balance of the evidence seems to lie firmly with Leonelli when she writes, directly responding to Anderson's editorial, that, "as recognized by both champions and critics of data-intensive methods, extracting biologically meaningful inferences from data involves a complex interplay of components, such as reliance on background knowledge, model-based reasoning, and iterative tinkering with materials and instruments" (Leonelli 2016, p. 135).

But the story isn't quite over, because we have only eliminated one of the two ways, mentioned above, in which my account is somewhat ambiguous. I have argued that investigating the possibility of theory-free biology confirms my assertion that integration should be interpreted as a property of scientific explanations, in order to avoid begging the question. But I have yet to clarify the sense of "integration" that is at work in these discussions of big data. Such a clarification is the final prerequisite for an overall evaluation of theory-free biology, which will be the subject of this chapter's final section.

6.2 Integration and "Data Integration"

As we have already noted, the word "integration" is found throughout work on data-driven life science. Leonelli puts the matter bluntly when she notes that "the main motivation behind the development of bio-ontologies is to facilitate data integration to foster further research" (Leonelli 2016, p. 123). But this invocation of *data* integration nicely instantiates a confusion that has run throughout this chapter, and which it is high time to resolve. When we talk about data integration in the context of data-centric biology, are we talking about the same sort of inter-domain integration that I detailed and defended in Chap. 2, or another sort of integration? If the latter, how should we understand the relationship between the two?

Pulling apart the threads of traditional integration and data integration is frequently difficult, in part because philosophical work on big data biology treats it not as a rejection of the theoretical tools of previous approaches within the life sciences, but rather something more like "a particular model of attention within research, within which concerns around data handling take precedence over theoretical questions around the logical implications of [a] given axiom or stipulation" (Leonelli 2016, p. 178). On this kind of view, we shouldn't expect straightforward incompatibilities between "traditional" and "big-data" biological projects, but instead the expression of a relative significance or fundamentality claim, the implications of which may be harder to describe. More concretely, it's not as though big-data biology must reject traditional inter-domain integration; it might rather consider it less important or less basic than some other sense of the term.

In this section, I argue that, indeed, data integration and inter-domain integration are two separate kinds of integration, occasionally conflated thanks to the ambiguous use of "integration" in the two cases. Contrary to arguments made by a number of philosophers that data integration and inter-domain integration are fundamentally linked, I will maintain that keeping these two types of integration distinct from one another is important not only to evaluate arguments for and against theory-free biology, but more generally to gauge the integrative promise of any data-driven research program.

6.2.1 From Data Integration to Inter-Domain Integration

First, what exactly do we mean by "data integration"?[7] At its simplest, we might be talking about one of the fairly regular steps on a data journey. Once scientists recognize that data "need to travel beyond the boundaries of their own investigation," one of their standard activities is to decide how their data can be "integrated with other data sources" (Leonelli 2015, p. 816). As we'll see when I present an example below, while this can require that scientists make some very important choices about how to gather and package their data, this is what we might call a "deflationary" reading of data integration: it's nothing more than the process of providing the relevant kinds of interoperability such that the data sources generated by one laboratory can be connected with those generated by colleagues.

A deflationary picture of data integration would not require any significant changes to the framework of integration as I have laid it out in Chap. 2. Data integration would, on this interpretation, be a *supplementary* method—sometimes, perhaps, quite significant as a way in which to enable inter-domain integration, but not a process that radically modifies what it means to be engaging in inter-domain integration, even in cases of big-data biology.

An "inflationary" reading, on the other hand, would make "data integration" a core element of "integration" in the broader sense that I've used throughout the book. For O'Malley and Soyer, for instance, the term "integration"

> is used to cover a multi-faceted dynamic, in which methods, bodies of data and explanations are synthesized in order to understand and intervene more effectively in biological systems. Integration can also include the combination of [data-driven] approaches with [hypothesis-driven] approaches, and emerging approaches to molecular biology often require this. Systems biology, the successor science to standard [data-driven] genomics, is usually described as the field that has arisen out of a perceived need for more integration, particularly of data and methods but also of disciplines. (O'Malley and Soyer 2012, p. 59)

For these authors, then, data integration is a part of inter-domain integration itself, which could then refer not only to the integration of explanations (as I've done here), but also to the integration of methodologies, data, and more. If an inflationary view

[7] For a representative view from the sciences, see the survey of David Searls (2005).

of data integration is true, then we are perhaps due to modify the framework that I've developed, giving data integration (as well as, potentially, other types besides) a more central role.

As I've already hinted at above, I think that a deflationary reading is the more justified approach to data integration. I'll make the case in two different ways. First, I'll consider O'Malley and Soyer's approach to data integration (as well as a third type of integration that they discuss, methodological integration). Then, I'll turn to an example of data integration, drawn from work by Oliver Lean (2021), which will make clearer why I want to stick with inter-domain integration as the primary sense relevant for integrative promise.

Largely driven by a desire to understand the nature and function of multi-level explanations in systems biology, O'Malley and Soyer argue for the importance of at least three senses of integration, all of which should, they claim, be simultaneously analyzed in understanding systems-biological work. First is *methodological* integration. They describe this as "directing a range of methods at a particular biological system or research problem in order to gain a multidimensional understanding of how the system works," where here understanding "means not just explanation, but also (very importantly for systems biology) prediction and control" (O'Malley and Soyer 2012, p. 60). Next comes data integration, "the activity of making comparable different data types from a huge variety of potentially inconsistent sources" (O'Malley and Soyer 2012, p. 61). And lastly comes explanatory integration, which includes both "the synthesis of different explanations as well as the import of explanatory (and predictive) models from other research into a specified domain of inquiry" (O'Malley and Soyer 2012, p. 61). They also note further that even this taxonomy likely misses a number of important senses of integration, including, at least, integration between disciplines.

How should we understand these three senses of integration—and more specifically, should they replace the idea of inter-domain integration in integrative promise? I think not. First, their notion of explanatory integration is more limited than what I've described as inter-domain integration. Importing models from another domain does not necessarily mean that we will arrive at an explanation that cuts across domains in the ways that we've seen throughout the examples in the first half of the book. Furthermore, too much hinges here on what we mean by "synthesis." If "the synthesis of different explanations" is taken to mean the production of "synthesized" explanations that go above and beyond the scope of the individual explanations taken separately, then such cases might count as examples of inter-domain integration. But if we just mean something weaker, a kind of loose confederation of explanations, each of which offers limited, practical insight into the functioning of a part of a system, then it doesn't seem that integrative promise is genuinely being invoked.

The same is the case for both methodological and data integration. To be sure, being able to approach a system using methods or data drawn from different disciplines might be crucial to assuring the integrative promise of a putative explanation. As we saw in the case of EMBL in Chap. 4, inter-domain integration will certainly sometimes require the integration of methodologies. Their "planetary biology" theme incorporated insights from genetics and genomics, on through cellular and developmental biology to ecology. These disciplines clearly bring with them their own, independent methods. And the same is true for their data: we would need to ensure that genomic databases can interoperate with cellular or developmental databases, and that insights drawn from these could be applied to existing corpora of populational or ecological data.

But precisely the point is that EMBL does not stop there. We do not see claims that the integrative promise of EMBL's planetary biology program derives directly from its being able to use more than one method at once, or from its integrating disparate biological datasets. These tools are, rather, means to an end: grand challenges will be resolved by our ability to explain and understand the natural world. And that explanation and understanding moves beyond the ability to make methodologies or datasets talk to one another.

O'Malley and Soyer do have a number of arguments against the kind of view for which I've just argued. First, they write,

> grander unifying senses of integration are not particularly good descriptions of day-to-day instances of the practical application of explanatory integration, in part because models, explanations and theoretical resources are always contextualized, and connections are made for practical rather than theoretical reasons. (O'Malley and Soyer 2012, p. 62)

Here, I think, we see yet another reason for which I've underlined the role of stringency and opportunism in integrative promise. They are quite right that we must take into account the contextualization of explanatory resources in order to understand integrative promise—but this is precisely the role that, I've argued, stringency and opportunism play in these discussions. The practical and theoretical underpinnings that enable a particular explanation to exhibit integrative promise are, indeed, often front and center, and the way in which I've constructed integrative promise thus far makes room for just that emphasis.

Lastly, before turning to several examples of integration in systems biology, O'Malley and Soyer write that they consider those examples to "show the necessity of considering integration more inclusively than is captured in purely theoretical explanatory integration (as if methods and data follow sheepishly behind theory)" (O'Malley and Soyer 2012, p. 62). Again, I think that their conclusion is too quick. To say that methodological or data integration might be key prerequisites for the possibility of inter-domain integration in my sense is not at all to say that methods and data follow sheepishly behind theory. Indeed, as integrative promise is forward looking, it can serve precisely to highlight the weaknesses in our extant tools, weaknesses that could threaten our ability to carry through on an explanation's integrative promise. (Think, for instance, of the user scenarios for iPlant mentioned above, or EMBL's note that one of their grand-challenge programs would require developing a currently absent specialty in botany.)

To sum up, then, I don't think that any of the arguments offered by O'Malley and Soyer should convince us to alter the account of inter-domain integration that I've been using thus far. But that doesn't yet amount to an argument in favor of a deflationist reading of data integration. For that, we'll need to see a case study: let's look at the use of ontologies in biological data management.

6.2.2 Integrating with Ontologies

An ontology, in the sense that the term is used in data science, is not to be confused with the branch of metaphysics treating the philosophical study of being—although one can readily see why the term was borrowed. A data ontology categorizes, classifies, and taxonomizes a dataset, giving us a standardized vocabulary for talking about what kinds of entities and relations the dataset contains. As a radically over-simplified example, in a dataset describing observational embryological data, we would need terms in our ontology referring to specific samples, to species, and to the organs whose measures we've taken. We also need terms describing relations between those things: that particular measures of organs were taken from particular specimens, or that specimens instantiate species.

The overarching goal of data ontologies, then, is to explicitly encode all of this information in the database, in a way that can be "understood" both by human data curators and automated analysis tools. This is taken to serve at least three purposes. First, data that comes with an ontology is easier for scholars who were not involved in its production to interpret. If we have fully specified our ontology, it's much less likely that a published dataset will fail to mention, for instance, what units were used to measure the length of an organ, or the growth stage of an embryo when it was measured. Making these features of the dataset explicit serves as a kind of self-check that allows data using ontologies to, at least potentially, be of higher quality.

Second, data that uses ontologies, by expressing the relationships between data in a manner that can be accessed computationally, enables automated reasoning on those bodies of data. If I obtain one dataset that describes the sizes of organs in terms of growth stages of an embryo, and another dataset that describes gene expression at those same growth stages in those same organs—where here, the "same" growth stage and organ are explicitly defined by the ontology—then I could immediately hunt for correlations between gene expression and organ growth. Moreover, I can do that without having to have a research assistant manually verify that, for instance, both laboratories were using the same definitions of each organ, as the ontology provides an external standard.

Finally, and most importantly for us, by offering just such an external standard—and as we saw in the example that I just sketched—data ontologies are taken to make data integration much easier. As we already saw in Leonelli's view of data journeys, internal choices about standardization and data curation are essentially always taken in the practice of data-centric biology. Ontologies make those choices externally visible and, if a community standard can be built around them, serve as a way to unify laboratories working on a particular question, massively facilitating the future task of data integration. An external, public standard can, at least potentially, become a community norm.

Consider a real example in Fig. 6.2. This is a portion of the Gene Ontology (GO), the leading ontology for describing entities and concepts in molecular biology (Ashburner et al. 2000; The Gene Ontology Consortium 2021). At the center, we have the term "nucleus." The visualization has filtered out all other concepts

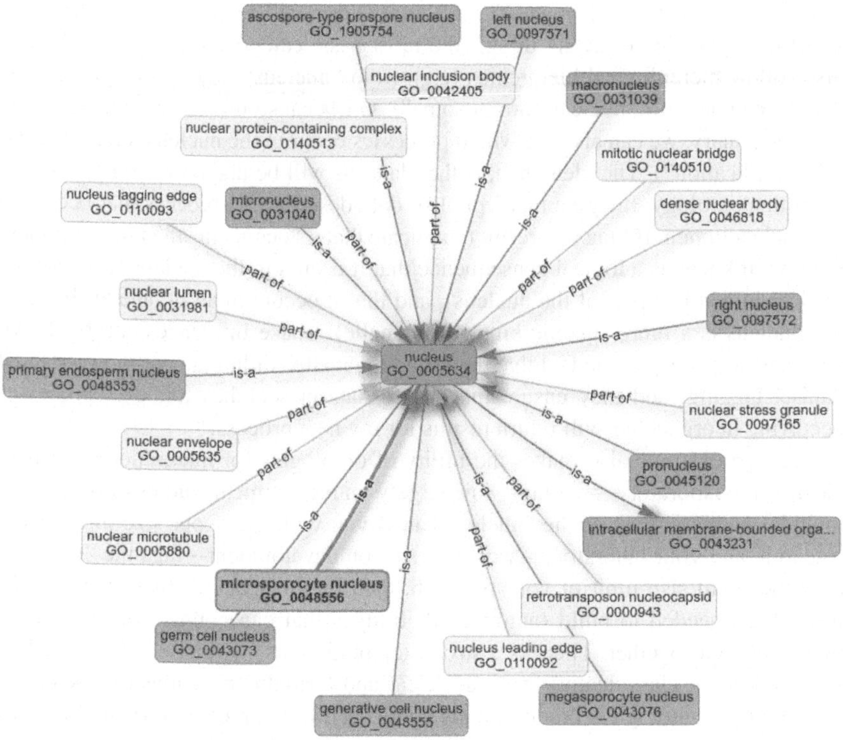

Fig. 6.2 Screenshot of the network of terms and relations (restricted here to "is a" and "part of") that point to the term "nucleus." Figure produced by EMBL-EBI's Ontology Lookup Service (OLS) (Jupp et al. 2015) with data drawn from the Gene Ontology (GO; Ashburner et al. 2000; The Gene Ontology Consortium 2021), CC-BY 4.0

except those connected to "nucleus" by two relations: "*X* is a *Y*" (in the sense of instantiating a more general class) and "*X* is part of *Y*" (in the sense of mereological parthood). Thus, we have in this image all the concepts in the ontology which, in general, *are* a nucleus (in the sense of being a more specific kind of nucleus), as well as those concepts in the ontology that are *part of* a nucleus. The network here thus tells us that there are twelve more specific kinds of nucleus in the Gene Ontology (e.g., the microsporocyte nucleus), and that the GO subdivides the nucleus into eleven general kinds of parts (most of which, in turn, are subdivided further by additional "part of" relationships). To maximize the ease of automated inference from ontologies, the relations encoded in an ontology are often limited to only these two, though the Gene Ontology adds to them, for instance, the relation "occurs in" as a way to localize not just objects but biological functions within a particular part of the cell.

One can therefore use the ontology to trace out relationships in a number of senses. We can follow specificity of description (e.g., from "nucleus" to the more precise "germ cell nucleus" and "male germ-cell nucleus", or to the more general "organelle" or even "thing," the most basic category for objects). We can also follow mereological hierarchies (e.g., from "nucleus" down to its proper part "nuclear lumen" to "nuclear chromosome"). Or, because the Gene Ontology adds this vocabulary, we can also see which processes occur in the nucleus (e.g., "nuclear DNA replication"). The idea, then, is that datasets will be tagged with the concepts from this ontology. Imagine that I publish data describing a process that occurs in the nuclear lumen. If I tag that result to indicate that it "occurs in" the nuclear lumen, then we'll know as a logical consequence that it occurs in the nucleus (because the nuclear lumen is a part of the nucleus), and that it occurs in an organelle (because the nucleus is a more specific kind of organelle). These inferences can be drawn automatically, despite the fact that the term "organelle" didn't appear in my manual dataset tagging, and they ensure that future dataset searches for, e.g., processes occurring in organelles will return results for my new process.[8]

Such an example also makes the utility of ontologies for integration extremely clear. If my laboratory—working, say, largely on biochemical studies of proteins—includes some tagging of the "nucleus" at, if you will, the upper size-limit of our research, and your laboratory—working, say, on development—tags the "nucleus" near the small size-limit of your research, then the presence of these tags can be just what's needed to build (automated!) bridges that can connect our respective bodies of data together. There is a strong temptation, as Oliver Lean has noted, to see this kind of classificatory work as "clear and formal expressions of a scientific community's ontological commitments about their domain of inquiry, and to view their integration as tantamount to the metaphysical unification of science that some philosophers have envisaged" (Lean 2021, p. 11587). If we can really all place our metaphysical posits on the "common ground" of the Gene Ontology, for instance, then aren't O'Malley and Soyer right that there is a deep connection between the practice of data integration and inter-domain integration more broadly?

I think Lean offers us two good reasons to be skeptical. First, data integration via ontologies does not automatically entail the possibility of inter-domain integration, in part because ontologies often "need to account for the diversity and dissensus that exists between the participating communities" building and using an ontology (Lean 2021, p. 11598). As Leonelli puts the point, ontologies

> need not be universal; rather, they capture the assumptions, interpretations, and practices underlying the successful sharing and reuse of data within very specific contexts at a given moment in time and provide a long-term methodological vision for how society should go about investigating the natural world. (Leonelli 2016, p. 135)

This gives them a kind of paradoxical role: acting both as guarantors of integration and carrying a long-term vision, they nonetheless reflect local, short-term, situated decisions about how data should be organized and how future investigations should

[8] Of course, these simple examples don't really touch on the power of such automated inferences, which can even infer things like the structure and function of as-yet undescribed proteins by drawing upon data about homologous DNA sequences that have been tagged in the ontology.

be performed. But this is only surprising, I think, if we have an inflationary reading of data integration. A deflationary conception, which puts less weight on data integration as part and parcel of a single process that also involves inter-domain integration, can easily accommodate this kind of conflict between the local and global character of ontologies. Ontologies' pretension to global scope will rest, ultimately, on our ability to construct explanations that utilize these ontologies—to actually build out inter-domain integrations that transcend the temporary and situated reasons for which ontologies were created by the particular scientific communities that developed them. We have, in turn, no guarantee that we'll ever get to those explanations, in the end. We don't need, and shouldn't expect, an argument that they are available in general, or, put differently, that there is any *a priori* connection between data integration and inter-domain integration.

Further, Lean notes that, as has been argued by Sterner et al. (2020), data integration of any sort isn't a necessary condition for the inter-domain integration of two data-driven areas of science. We can, and sometimes do, find ways in which to "pool data for integration purposes while allowing them to retain the idiosyncrasies of their origins, and to coordinate data representation in ways that honour dissensus in the definitions of terms" (Lean 2021, p. 11598). While one might read one's metaphysics directly from the content of a data ontology, arguing that such an ontology represents a "global consensus metaphysical interpretation of the reality those data describe" (Sterner et al. 2020, p. 6), this is not the only imaginable approach. Sterner et al. (2020, p. 10) note that some ontologies even relax the standards usually set on ontologies for the purposes of formal reasoning—the *Drosophila* anatomy ontology, for instance, allows for visual as well as textual definitions. On this kind of view, rather than reflecting an unambiguous underlying reality, coordination is maintained because ontologies are "grounded in a local consensus on principles for evaluating the quality of a definition" (Sterner et al. 2020, p. 10). Such local consensus might allow us to arrive at integrated explanations for phenomena—that is, to arrive at inter-domain integration—without having the kind of universal, unambiguous data integration often called for by ontology designers.

Summing up, it seems that we both occasionally integrate our data (in the minimal sense of combining them under a single overarching ontology) without proposing an inter-domain integration, and we occasionally arrive at inter-domain integrations in data-driven fields without integrating our data. An inflationary reading of data integration seems, in the end, to be unjustified.

6.3 Doing Without Theory?

It's time to return to a clearer picture of the prospects for theory-free biology. Of course, it should be no surprise at this point that my final evaluation will be largely negative—the weight of evidence is undeniable that the use of big-data methods requires, just as much as past work in the life sciences, a deep familiarity with the details of one's subject matter and a healthy dose of theoretical commitments as we manage data journeys, curate ontologies, and so forth. What I hope to show here is rather that an approach to this question in terms of integrative promise—of both theory-free and, to borrow Leonelli's coinage, "theory-informed" big-data biology—can offer us a number of advantages. First, we can construct a more reasonable position on behalf of theory-free biology, one that might let us understand why contemporary biologists seem to see it as a threat. Second, we can zero in on the debate over theory-free biology as one that concerns precisely the integrative promise of these competing classes of explanations for the same kinds of phenomena. Many of the arguments offered against theory-free biology follow exactly the form that one would expect in the context of an integrative-promise debate.

How does an integrative-promise reconstruction of the fight over theory-free biology go? Putting aside, for the moment, the heated rhetoric that big-data science means we no longer require theory or hypothesis, we can see a way in which to reconstruct a narrower debate about the integrative promise of two different kinds of explanations in data-driven science. On one hand, we have the "theory-free" narrative. Such an enterprise would consist in explaining (whether in a unifying or a pragmatic sense of the term) on the basis of correlations drawn from large corpora of biological data by digital analyses. In line with our discussion in the last section, this would not be merely the result of the data integration that has taken place—again, *contra* Anderson, it is not simply moving to the "petabyte scale" that gives rise to inter-domain integration. But on at least some understandings of explanation, it remains possible to envisage explanations that are drawn directly from correlations and that bring together domains either across the tree of life or at multiple levels of biological organization. Theory-free biology thus claims to be integrative in my sense.

It also seems to invoke both opportunism—clearly enough, given its reliance upon datasets that have, in many cases, already been collected—and stringency. The case for the latter merits a bit of further scrutiny. In the background of some of the appeals for theory-free biology seem to lie two observations that both appeal to the stringency of putative theory-free explanations. First, there is a kind of faith that in the context of a large enough dataset, spurious correlations will necessarily be driven out and signal will dominate over noise. In that sense, an explanation that can withstand being tested by the kinds of big data that we now possess is taken to have a kind of intrinsic stringency. Second, defenders of theory-free biology often appeal to the complexity of contemporary biological phenomena. If the search for mechanisms involves a necessary kind of simplification or black-boxing, then perhaps theory-free biology's insistence on collecting all potentially relevant data—on embracing that complexity rather than idealizing it—makes theory-free explanations more stringent.

Lastly, while we haven't seen much evidence of this in my presentation so far, theory-free science advocates do indeed appeal to the importance of their work for resolving the grand challenges of our day. In a wide-ranging work dedicated to the "fourth paradigm" of data-intensive science (and in many cases fully embracing a "theory-free" perspective), Microsoft computer scientist Dan Fay, immediately following the claim that we need "to better capture, analyze, model, and visualize scientific information," adds that

> As society demands action and responsiveness to growing environmental issues, new types of applications grounded in scientific research will need to move from raw discovery and eliciting basic data that leads to knowledge to informing practical decisions. Active issues such as climate change will not wait until scientists have all the data to fill their knowledge gaps. (Hey et al. 2009, p. 3)

The book as a whole is divided into grand-challenge-style sections entitled, among others, "Earth and Environment" and "Health and Wellbeing." The idea running throughout seems to be that the very same complexity mentioned above as making theory-free biology's explanations stringent tests of integration means also that theory-free biology will be a tool particularly well suited for attacking grand challenges that entail this same kind of complexity.

It thus seems fairly straightforward to build an integrative-promise defense of theory-free biology. Against this backdrop, how do the critiques that we saw earlier fare? First, we have arguments like those offered by Callebaut, who repudiated the idea that theory-free biology could get by without theories or models in complex domains. Such a rejoinder, in my vocabulary, seems to attack the idea that the explanations offered by theory-free biology are truly integrative. If what is required to tie together phenomena in a genuinely explanatory way is something much like a theory or model, then a theory-free biology project can't get off the ground. Whether this is because Callebaut rejects outright the views of explanation on which theory-free biology might be explanatory or because he doesn't believe that token theory-free, correlative explanations adequately explain phenomena in the examples he evaluates from systems biology and evolutionary developmental biology is less clear, but the point remains: for Callebaut, theory-free biology cannot be integrative.

Above, I noted that O'Malley et al. seem to be invoking stringency when they talk about "a judicious choice of physical entities and interactions" in the context of integration (O'Malley et al. 2014, p. 820). As we saw in the last section, we have to be at least somewhat careful here. The larger context of their discussion indicates that they are talking about the characteristics of scientific explanations, and hence we can understand them as arguing in favor of the stringency of theory-informed, data-centric biology. But this judicious choice-making also happens in data integration, which is not immediately related to questions of inter-domain integration. Keeping the two senses separate is important for just what kind of stringency we have in mind, and how these pragmatic considerations interact either with a deflationary sense of data integration or with the broader idea of inter-domain integration.

Lastly, I should turn to the role of integration in Leonelli's arguments against theory-free biology. When she claims that "the main motivation behind the development of bio-ontologies is to facilitate data integration to foster further research" (Leonelli 2016, p. 123), she is surely right—for the narrower sense of data integration. She then argues, building on this observation, that data ontologies function as a kind of "classificatory theory" for data-driven science. The relationship between that argument and my project here needs to be spelled out in more detail, and that nuance will resolve what I had marked out earlier as a potential conflict between her view and mine. It is at precisely this stage of the discussion that Leonelli makes the case that integration, when it occurs in data-driven biology, is a property of theories (*sensu* Mary Hesse) rather than of "detailed explanations or daring new hypotheses" (Leonelli 2016, p. 124). But if we see the notion of integration at play for Leonelli here as data integration rather than inter-domain integration, then the idea that ontologies-as-theories are crucial for data integration is entirely consonant with my position that explanations play the leading role in inter-domain integration. Again, we've been bitten by the polysemy of "integration"—and separating out data integration from the rest lets us see where we were led astray.

The overall critique that Leonelli offers of theory-free biology, then, takes on two different aspects. First, she argues that practical and experimental know-how are vital for discerning whether a particular dataset has evidential value in a given context—that is, whether or not the employment of a dataset can possibly serve as a stringent test of an integrative explanation. Size is not, *pace* defenders of theory-free biology, the only thing that matters. What's more, she offers a pointed argument about theory-free biology's ability to respond to grand challenges. We cannot say ahead of time whether a given big-data explanation will serve to resolve an especially socially or ethically important problem—in particular, not without foregrounding just those aspects of data journeys that advocates of theory-free biology so often set aside (Leonelli 2016, p. 164). Theory-free biology thus runs a serious risk of techno-solutionism, a kind of pseudo-resolving of grand challenges by well-financed, well-hyped research methods that leave the fundamental problem untouched.

On the one hand, this chapter has been a relatively circuitous route to a reasonably predictable conclusion. Big-data biology, not unlike other splashy technological advances, has offered overblown predictions of a theory-free future for the life sciences, predictions which have in turn come in for harsh philosophical critique. I've argued that deploying the resources developed in this volume for identifying integrative promise sheds light on this debate in at least two ways. First, it can help us see that the sense of "data integration" often discussed in the philosophical and scientific literature is best analyzed by not assimilating it to other notions of integration (like the concept of inter-domain integration as an explanatory virtue that I have developed here). This, then, lets us have a more sophisticated understanding of the criticisms that philosophers have launched against the very idea of theory-free biology—criticisms that, I think, still clearly carry the day, but that take on a new structure and importance when we see them as targeted at a debate over the integrative promise of theory-free biology versus alternatives like multi-scale modeling in systems biology.

References

Anderson C (2008) The end of theory: the data deluge makes the scientific method obsolete. Wired

Ashburner M, Ball CA, Blake JA, Botstein D, Butler H, Cherry JM, Davis AP, Dolinski K, Dwight SS, Eppig JT, Harris MA, Hill DP, Issel-Tarver L, Kasarskis A, Lewis S, Matese JC, Richardson JE, Ringwald M, Rubin GM, Sherlock G (2000) Gene ontology: tool for the unification of biology. Nat Genet 25(1):25–29. https://doi.org/10.1038/75556

Baker A (2005) Are there genuine mathematical explanations of physical phenomena? Mind 114(454):223–238

Bartholomew GA (1986) The role of natural history in contemporary biology. BioScience 36(5):324–329. https://doi.org/10.2307/1310237

Callebaut W (2012) Scientific perspectivism: a philosopher of science's response to the challenge of big data biology. Stud Hist Philos Biol Biomed Sci 43(1):69–80. https://doi.org/10.1016/j.shpsc.2011.10.007

Craver CF, Darden L (2013) In search of mechanisms: discoveries across the life sciences. University of Chicago Press, Chicago

Emerson RA, East EM (1913) The inheritance of quantitative characters in maize. No. 2 in Bulletin of the Agricultural Experiment Station of Nebraska, University of Nebraska, Lincoln

Glennan SS (2002) Rethinking mechanistic eplanation. Philos Sci 69(3):S342–S353

Goff S, Vaughn M, McKay S, Lyons E, Stapleton A, Gessler D, Matasci N, Wang L, Hanlon M, Lenards A, Muir A, Merchant N, Lowry S, Mock S, Helmke M, Kubach A, Narro M, Hopkins N, Micklos D, Hilgert U, Gonzales M, Jordan C, Skidmore E, Dooley R, Cazes J, McLay R, Lu Z, Pasternak S, Koesterke L, Piel W, Grene R, Noutsos C, Gendler K, Feng X, Tang C, Lent M, Kim Sj, Kvilekval K, Manjunath B, Tannen V, Stamatakis A, Sanderson M, Welch S, Cranston K, Soltis P, Soltis D, O'Meara B, Ane C, Brutnell T, Kleibenstein D, White J, Leebens-Mack J, Donoghue M, Spalding E, Vision T, Myers C, Lowenthal D, Enquist B, Boyle B, Akoglu A, Andrews G, Ram S, Ware D, Stein L, Stanzione D (2011) The iPlant collaborative: cyberinfrastructure for plant biology. Fronti Plant Sci 2. https://doi.org/10.3389/fpls.2011.00034

Green S (2022) Philosophy of systems and synthetic biology. https://plato.stanford.edu/archives/sum2022/entries/systems-synthetic-biology/

Hesse MB (1980) Revolutions and reconstructions in the philosophy of science. Harvester, Brighton

Hey T, Tansley S, Tolle K (eds) (2009) The fourth paradigm: data-intensive scientific discovery. Microsoft Research, Redmond

Jupp S, Burdett T, Malone J, Leroy C, Pearce M, McMurry J, Parkinson H (2015) A new ontology lookup service at EMBL-EBI. In: Malone J, Stevens R, Forsberg K, Splendiani A (eds) Proceedings of SWAT4LS international conference 2015, Cambridge

Kitcher P (1981) Explanatory unification. Philos Sci 48(4):507–531

Lean OM (2021) Are bio-ontologies metaphysical theories? Synthese 199(3):11587–11608. https://doi.org/10.1007/s11229-021-03303-4

Ledgerwood A (2018) The preregistration revolution needs to distinguish between predictions and analyses. Proc Natl Acad Sci 115(45):E10516–E10517. https://doi.org/10.1073/pnas.1812592115

Leonelli S (2015) What counts as scientific data? A relational framework. Philos Sci 82(5):810–821. https://doi.org/10.1086/684083

Leonelli S (2016) Data-centric biology: a philosophical study. University of Chicago Press, Chicago

Machamer P, Darden L, Craver CF (2000) Thinking about mechanisms. Philos Sci 67(1):1–25. https://doi.org/10.1086/392759

Merchant N, Lyons E, Goff S, Vaughn M, Ware D, Micklos D, Antin P (2016) The iPlant Collaborative: cyberinfrastructure for enabling data to discovery for the life sciences. PLoS Biol 14(1):e1002342. https://doi.org/10.1371/journal.pbio.1002342

Needham J (1932) Thoughts on the problem of biological organization. Scientia 52:84–92
Nicholson DJ, Gawne R (2015) Neither logical empiricism nor vitalism, but organicism: what the philosophy of biology was. Hist Philos Life Sci 37(4):345–381. https://doi.org/10.1007/s40656-015-0085-7
Nosek BA, Ebersole CB, DeHaven AC, Mellor DT (2018) The preregistration revolution. Proc Natl Acad Sci 115(11):2600–2606. https://doi.org/10.1073/pnas.1708274114
O'Malley MA, Soyer OS (2012) The roles of integration in molecular systems biology. Stud Hist Philos Biol Biomed Sci 43(1):58–68. https://doi.org/10.1016/j.shpsc.2011.10.006
O'Malley MA, Brigandt I, Love AC, Crawford JW, Gilbert JA, Knight R, Mitchell SD, Rohwer F (2014) Multilevel research strategies and biological systems. Philos Sci 81(5):811–828. https://doi.org/10.1086/677889
Pearson K (1894) Dilettantism and statistics: [review of] The Growth of St. Louis Children. Nature 51(1311):145–146. https://doi.org/10.1038/051145a0
Popper KR (1983) Realism and the aim of science. Routledge, London
Salmon WC (1984) Scientific explanation and the causal structure of the world. Princeton University Press, Princeton
Sarkar S (2023) That was the philosophy of biology that was: Mainx, Woodger, Nagel, and logical empiricism, 1929–1961. Biol Theory. https://doi.org/10.1007/s13752-023-00429-1
Schindler S (2018) Theoretical virtues in science: uncovering reality through theory. Cambridge University Press, Cambridge
Searls DB (2005) Data integration: challenges for drug discovery. Nat Rev Drug Discov 4(1):45–58. https://doi.org/10.1038/nrd1608
Skow B (2014) Are there non-causal explanations (of particular events)? Br J Philos Sci 65(3):445–467. https://doi.org/10.1093/bjps/axs047
Stern CD (2022) Reflections on the past, present and future of developmental biology. Dev Biol 488:30–34. https://doi.org/10.1016/j.ydbio.2022.05.001
Sterner B, Witteveen J, Franz N (2020) Coordinating dissent as an alternative to consensus classification: insights from systematics for bio-ontologies. Hist Philos Life Sci 42(1):8. https://doi.org/10.1007/s40656-020-0300-z
The Gene Ontology Consortium (2021) The Gene Ontology resource: enriching a GOld mine. Nucleic Acids Res 49(D1):D325–D334. https://doi.org/10.1093/nar/gkaa1113
van Fraassen BC (1977) The pragmatics of explanation. Am Philos Q 14(2):143–150
van Ommen GJB (2008) Popper revisited: GWAS here, last year. Eur J Human Genet 16(1):1–2. https://doi.org/10.1038/sj.ejhg.5201970
Venter JC, Remington K, Heidelberg JF, Halpern AL, Rusch D, Eisen JA, Wu D, Paulsen I, Nelson KE, Nelson W, Fouts DE, Levy S, Knap AH, Lomas MW, Nealson K, White O, Peterson J, Hoffman J, Parsons R, Baden-Tillson H, Pfannkoch C, Rogers YH, Smith HO (2004) Environmental genome shotgun sequencing of the Sargasso Sea. Science 304(5667):66–74. https://doi.org/10.1126/science.1093857
von Bertalanffy L (1933) Modern theories of development: an introduction to theoretical biology. Oxford University Press, London
Weldon WFR, Pearson K, Davenport CB (1901) Editorial: the scope of Biometrika. Biometrika 1(1):1–2. https://doi.org/10.1093/biomet/1.1.1
Wolfram S (2021) Multicomputation: a fourth paradigm for theoretical science. https://writings.stephenwolfram.com/2021/09/multicomputation-a-fourth-paradigm-for-theoretical-science/
Woodger JH (1929) Biological principles. Kegan Paul, London

Chapter 7
The Promise of Model Organism Research

Abstract Research with model organisms is often lauded for its putative ability to integrate knowledge at the molecular, physiological, and organismic levels, and across the tree of life. I'll explore this *prima facie* surprising assertion (why would working on a single organism better enable us to integrate across multiple levels and multiple organisms?), and then introduce a case study describing the use of ferrets in research in virology. Ferrets, which can be understood as a kind of "model organism in development," clearly exhibit the different ways that scientists both work to enhance and improve the integrative promise of their work over time and deploy arguments that this integrative promise merits the support of the broader scientific community. I'll both describe the current state of play in such research and discuss the calls for further enhancement and refinement of the ferret model in order to make the case.

Keywords Model organisms · Integration · Databases · Virology · Ferret · Experimental organisms

If, as we saw in the last chapter, big-data biology is among the places where we find that talk of integration is commonplace, assuredly another such domain is that of model organism research. No less central to the practice of contemporary biology than data-driven methodologies, the codification of an explicit list of model organisms by the U. S. National Institutes of Health in 1999 has been taken by many to be a watershed moment, recognizing the importance of results drawn from a relatively limited set of species for contemporary biomedical research. Unfortunately, the original resource from the NIH is no longer available, but Rachel Ankeny and Sabina Leonelli (2020, p. 2) note that the list contained thirteen species in total, including all of the expected "heavy hitters": the mouse (*Mus musculus*), the rat (*Rattus norvegicus*), the fruit fly (*Drosophila melanogaster*), the nematode (*Carnorhabditis elegans*), yeast (*Saccharomyces cerevisiae*), the zebrafish (especially important in developmental biology research, *Danio rerio*), and mouse-ear cress (the primary botanical model organism, *Arabidopsis thaliana*). As Ankeny has argued, these organisms present vitally important loci for philosophical study,

C. H. Pence, *Integrative Promise*, Synthese Library 499,
https://doi.org/10.1007/978-3-031-88471-9_7

if only because they "garner most of the attention of researchers and dictate the distribution of most of the biomedical research funds that are not targeted to specific diseases" (Ankeny 2010, p. 95).

Indeed, there is now a robust literature surrounding model organisms—in particular, concerning the question of how exactly they serve as *models*. Model organisms, just as other kinds of scientific models, are understood to "stand in for" a broader class of organisms, and thereby to teach us about more general processes taking place across the tree of life. It is self-evident from the history of twentieth-century biology that they have successfully done so. To take just two examples from Arnon Levy and Adrian Currie (2015, p. 328), much of the early development of molecular genetics traces to discoveries in *Drosophila*, and work in mice became crucial for much of our knowledge of cancer and other human diseases.[1] Model organisms are thus a widely used and successful research tool.

In this chapter, I don't want to focus on the epistemic foundations of the knowledge that model organisms generate (interested readers can consult, among many others, Ankeny 2010; Leonelli and Ankeny 2012; Levy and Currie 2015; Ankeny and Leonelli 2020). I rather want to examine another facet of model organism work that is also regularly emphasized by philosophers and scientists alike: its integrative function. Model organism research is pursued, in part, because it is taken to let us integrate biological knowledge. At first blush, this might seem surprising: how could it be that using a smaller number of species allows us to integrate? But, as I will show in the next section, integrate they do, both horizontally and vertically. What, then, is the relationship between model organism research and integrative promise? After building up the theoretical case for integration using model organisms, I will turn to a case study of what we might call a "model organism in development": the use of the ferret for the study of infectious respiratory disease transmission and virulence. Debates over the quality of ferret-model results and over how we ought to advance the ferret model in the future will, I contend, come down once again to targeted concerns about the integrative promise of the program. Virologists committed to ferret-based research believe that it exhibits extremely high integrative promise, and argue for the investment of time and money in refining their system—such that, to use a helpful turn of philosophical phrase that I will elaborate below, the ferret's status passes from that of an experimental organism to a genuine model organism—precisely because that work will allow them to capitalize on this promise.

[1] Mice are so central, in fact, that they have been subject to no small amount of critique, whether philosophical (Ankeny 2010) or parody (a Twitter account named "@justsaysinmice" spent years re-posting over-hyped scientific headlines like "McDonald's fries can cure baldness" with the addition "...in mice").

7.1 Model Organisms and Integration

Without wanting to too narrowly define what a model organism is—as this would almost certainly require taking a position on the question of the nature of model organism-derived knowledge that I'm setting aside for the purposes of this chapter—we can start with a general list of their advantages as laid out by Ankeny and Leonelli:

> For instance, they are typically easy to breed and maintain in large numbers under laboratory conditions. Model organism research characteristically involves the standardisation of the organism in question and the accumulation of knowledge and resources on the organism on a large scale. These resources include relevant networks, organism-focused conferences, stock centres, and cyberinfrastructures. Such research is done with a view to creating a platform for interdisciplinary integration across biological disciplines and a reference point for comparative research across species. (Ankeny and Leonelli 2020, p. 3)

This nicely highlights the two dimensions of integration that model organisms are normally taken to involve.[2] First, by creating a "platform for interdisciplinary integration," they enable vertically integrative efforts: scientists working in a variety of different biological sub-disciplines have a common goal and a common collection of resources, which can encourage useful sites for interdisciplinary collaboration. Second, by creating a standard against which other research can be compared, they allow for horizontal integration, enabling well-grounded research that can nonetheless span large portions of the tree of life. Let's consider each in turn.

7.1.1 Vertical Integration and Platforms

As has been compellingly argued by James Griesemer, while model-organism research might have historically involved the use of a single organism to evaluate a single phenomeon, however important (think of the tight link between molecular genetics and *Drosophila*), this is no longer the case, as

> in the later 20th century, "model organisms" have come to serve as platforms for a wider range of research problems, often pursued by distinct specialities, whose interaction is facilitated by their shared materials and "common knowledge." (Griesemer 2013, p. 526)

[2] Leonelli has also drawn links between this integrative capacity of model organism research and the development of digital databases, noting that "the expansion of community databases has shifted attention away from research on single species and toward comparative, cross-species research that takes account of environmental variability" (Leonelli 2016, p. 145). For more on the relationship between integration and big data, see Chap. 6.

I don't follow Griesemer's full account of integration, as he considers it to be a property of "approaches," which include not just explanations, but also "practices involving appropriate use of tools and instruments, standardized procedures and protocols, and, in general, conventional practices and common knowledge" (Griesemer 2013, p. 528). (For my argument against this view, see my discussion of the very similar sense of integration defended by O'Malley and Soyer (2012), in Sect. 6.2.1.) But he is quite right in noting that one of the crucial roles of model organisms has been to furnish these sorts of "points of reference" for interdisciplinary work. In his case study, salamander tongue research performed by David Wake and colleagues started as a fine-grained evolutionary and functional study, and then expanded to include contributions from interdisciplinary figures working on biomechanics, phylogeny, and convergent evolution, all made possible by the common "platform" of the salamander.

More precisely, Ankeny and Leonelli (2020, pp. 23–25) locate the strength of this vertical integration in one key fact about model organisms: the limitation of their variability. By tightly controlling their environments, conditions of growth, and so forth, model organisms allow for inferences that might otherwise depend on various features of environment, genetics, or nutrition to be reliably assembled and compared with one another. Given this limited variability, for instance, an in-vitro study of a protein in one particular laboratory is likely to deal with an object reasonably similar to the protein as it is found in an in-vivo study in another group. They refer to this as a process of abstraction: model organisms become, on this view, a different kind of entity, one for which many of the details that are normally vitally important for the study of life can be removed without threatening the epistemic credentials of the knowledge thereby produced.

But as they go on to note, the mere presence of a model organism does not guarantee this kind of integrative power. Rather, a whole "modelling ecosystem ... makes it possible for model organisms to function as reference points for biological integration across levels. In turn, the development and upkeep of this modelling ecosystem depends on the resourcing, organisation, and institutionalisation of research communities and practices" (Ankeny and Leonelli 2020, p. 31). The importance of this insight will become clearer in the case study that I develop below. But for now, it suffices to say that the argument that model organisms offer us a concretely demonstrated locus for vertically integrative work is well established in the philosophical literature. Processes of standardization of organisms, protocols, reagents, and so forth offer us one of the best guarantees we could ask for that results produced in different laboratories by different investigators coming from different disciplinary traditions will still, in the relevant sense for the construction of scientific knowledge, be *about* the same thing.

7.1.2 Horizontal Integration and Comparative Analyses

Paralleling Griesemer's observation above that the vertical-integration role of model organisms is itself a historically contingent construction, Ankeny and Leonelli close their short book by writing that these models "are in the process of being shifted yet again to a range of different roles. In particular, they are being used for comparative and integrative investigations that take the role of environment as something to be studied rather than held constant" (Ankeny and Leonelli 2020, p. 70). This certainly seems to be the case when we look at the language used to describe model organism research in the European Molecular Biology Laboratory's mission statement for the next four years (for a broader analysis of this document, see Chap. 4). EMBL argues that their future projects will likely need to develop new model organisms in order "to answer particular questions related to the environment," especially to "identify the cellular mechanisms that underpin responsiveness and the environmental factors that drive organism adaptations and ultimately their evolution" (European Molecular Biology Laboratory 2022, p. 113). These kinds of endeavors, they write, will offer "a crucial starting point for studying interactions within and between species, and for modeling ecosystems in the lab, to understand how they work" (European Molecular Biology Laboratory 2022, p. 113).

This function of comparison has long been discussed in the model-organism literature: Levy and Currie, for instance, noted that on their view, "model organism work is part of a distinctive class of biological inference strategies, known as the comparative method . . . [which] gain epistemic traction through ancestral relations" (Levy and Currie 2015, p. 337). Whether or not one follows them in arguing that this form of inference grounds *all* work on model organisms, we do immediately see a basis for at least one kind of horizontal integration that model organisms might facilitate. T. H. Morgan, as they put it, considered *Drosophila* to be a model appropriate for all molecular genetics on the basis of an inference that the relevant common ancestor between fruit flies and all other sexually-reproducing organisms would have shared enough of their fundamental genetic machinery that studying those flies is useful to teach us about this larger class (Levy and Currie 2015, pp. 338–339).

The basic idea, then, is once again quite simple. How ought we best perform comparative, horizontally-integrative biological studies? We could, of course, just attempt to collect as much data on as many species as possible, and only afterwards attempt to elicit patterns from that data (though see my skepticism on this approach in Chap. 6). Alternatively, and likely more promisingly, we could start from a kind of pre-indicated point of reference, attempting to understand both the nature of our reference system as well as causal explanations for deviations from that point of reference.[3] Those inferences could be justified in a variety of ways, according to a

[3] One is reminded of Tim Maudlin's argument that we tend to appreciate theories that have a "*quasi-Newtonian*" form, by which he means that there are inertial laws that describe the "default" state of a system when nothing acts upon it, then laws of deviation that describe how a system can leave that default state (Maudlin 2007, p. 156).

Fig. 7.1 A domesticated ferret named Roy. Wikimedia Commons, CC-BY-SA 4.0, by Alfredo Gutiérrez

variety of analogies, perhaps strongest among them being the idea that phylogenetic relationships assure a relevant kind of similarity between the model organism and our integrative target of study. Model organisms, because of the amount of data accumulated about them as a "reference," are particularly well suited to do so.

So much for the first part of the case: model organisms are viewed by many philosophers and biologists as playing an important vertically and horizontally integrative role in contemporary life-science research. This poses an obvious question: are model organism research programs another place where we might find invocations of integrative promise? In the rest of this chapter, I will argue that they are, by offering a case study: the use of the ferret (*Mustela furo*, Fig. 7.1) in respiratory disease research.

7.2 Gain of Function Research in Virology

Before we explore why the ferret model serves as a particularly good case-study for the integrative promise of model organism research, we should start with some background. Why is it that, of all the models that one might discuss, the ferret deserves special philosophical attention? The answer comes from the link between ferret model research and science policy, which has ensured that a significant amount of international attention has been dedicated to the ferret. In 2011, two experiments on influenza were announced for publication in *Nature* and *Science*, describing work by teams led by Ron Fouchier at Erasmus University in the

Netherlands, and Yoshihiro Kawaoka at the University of Wisconsin-Madison in the US, respectively (Herfst et al. 2012; Imai et al. 2012). What made these studies noteworthy was the controversy that surrounded them. Both described the generation of novel strains of influenza—one, a mutated version of HA H5N1 "avian influenza," the other a recombinant strain of flu created from both H5N1 and the 2009 H1N1 pandemic "swine flu" strain—that were transmissible between mammals (Evans 2013; Evans et al. 2015). The generation of H5N1 flu viruses that transmitted between mammals had never been accomplished, in lab or in nature. And while seasonal influenza has a case fatality rate of approximately 0.1%, H5N1 has killed 60% of the people who have contracted it, and it is estimated that it would have a case fatality rate of 14–33% if it ever caused a pandemic (Li et al. 2008).

The fracas emerged because experts and policymakers were divided as to the merits of conducting or publishing these "gain of function" (GOF) studies. Security-minded folk pointed out that the two experiments were first and foremost a blueprint for potential bioterrorists who wanted to create a devastating and indiscriminate weapon (Evans 2013). Over the following decade, the debate would expand to influenza and coronavirus research more broadly, and focus on the safety issues that would arise as these techniques proliferated across the world and repeated experiments threatened a laboratory accident seeding a global pandemic (Evans et al. 2015; Evans 2018; Gryphon Scientific 2016; Lipsitch 2018).

But the scientific community, while divided in certain corners, largely closed ranks around the science, asserting, in the first place, that they would not withhold the research, and, second, that such work not only had scientific but *social* value. The former claim came down to a defense of a collection of values about scientific freedom and the importance of norms of openness for scientific progress (Evans et al. 2022). The latter was more ambiguous, however. Proponents claimed that the study of flu in mammals—and in ferrets in particular—was essential to maintaining surveillance for potential disease pandemics in nature; was useful for discovering medical countermeasures against influenza; and could aid the development of vaccines (Casadevall et al. 2014; Casadevall and Imperiale 2014; Kilianski et al. 2016; Schultz-Cherry et al. 2014). These proponents, however, attempted to play down the security and safety fears surrounding GOF experiments by noting that the creation of novel strains of influenza that were transmissible between ferrets did not necessarily mean those same strains would transmit between humans (Enserink 2011).

This may strike the reader as scientists attempting to have their cake and eat it too. The strength of the connection between ferrets as an experimental organism (to anticipate a term that I will define below) and humans motivates research to determine which mutations in H5 viruses would lead to sustained mammalian transmission. But that connection is not strong enough, it is also claimed, to ground fears that the results of the studies in ferrets are directly transferable to humans. Rather than a mere rhetorical device, however, this tension points to deeper questions about flu studies, and about the role of ferrets within them.

Given the stakes of GOF research, a number of authors have called for a more careful analysis of the cost-benefit calculus surrounding GOF in influenza

(Casadevall et al. 2014; Gryphon Scientific 2016). It's further clear—a point occasionally recognized but not widely appreciated in the literature, about which more in the next section—that such a cost-benefit analysis needs to fully consider not only the practical or ethical stakes, but also the *epistemic* stakes. What kinds of scientific knowledge could really be generated by GOF research, and what would the cost of foregoing that knowledge actually be? Performing that analysis and taking the epistemic stakes seriously requires that we explore the system in which that knowledge is generated: the ferret.

As we've already seen, there is an extensive literature on model organisms—but it has not yet spent time on ferrets, either in a merely descriptive or in a more normative dimension, i.e., considering whether ferrets meet the criteria for a model-organism system in the traditional sense, or are rather what have been called "experimental organisms" or "Krogh organisms" (Green et al. 2018), which exhibit less standardization and a narrower inferential scope. I'm introducing the ferret case here precisely due to this ambiguity: given that ferrets seem to be a case of an organism *between* these two poles, we have what we might call a model organism in-the-making, or an experimental organism with aspirations to become a model organism. This will, in the last section, allow me to explore arguments about the usefulness of the ferret model as arguments over its integrative promise: integrative promise is indeed a feature of discussions about model organism research, and recognizing this can help us understand some of the otherwise confusing behavior of the scientists working on these studies.

7.3 The Ferret Model of Influenza

The ferret model of influenza infection and transmission has been linked with the study of influenza since the very first days of research into the virus. After an influenza epidemic in 1933, Wilson Smith and colleagues at the National Institute for Medical Research's Farm Laboratories in Mill Hill, outside of London, collected throat washings from a number of patients, and proceeded to assay every available experimental animal in their facility to see if any of them could become infected with the flu (Smith et al. 1933). By a stroke of luck, a population of experimental ferrets was being maintained to investigate canine distemper virus, and two ferrets exposed began to demonstrate all the classic symptoms of the flu: fever, food avoidance, weight loss, sneezing ("the ferret has an exquisite sneeze reflex," write Radigan et al. 2015, p. 315), fatigue, and a runny nose (Belser et al. 2016, p. 734). The work of Smith et al., in the end, led to the first-ever isolation of the influenza virus, which had been suspected but not confirmed to be the method of the illness's transmission.

What makes the ferret a particularly useful model for influenza research? Ferrets are, as it turns out, the only known organism which "can present both the pathogenic and transmissible features of influenza virus infection" (Belser et al. 2011, p. 575). That is, ferrets will both present a course of disease symptoms similar to humans (Fig. 7.2)—giving us an idea of whether or not a particular strain of influenza would

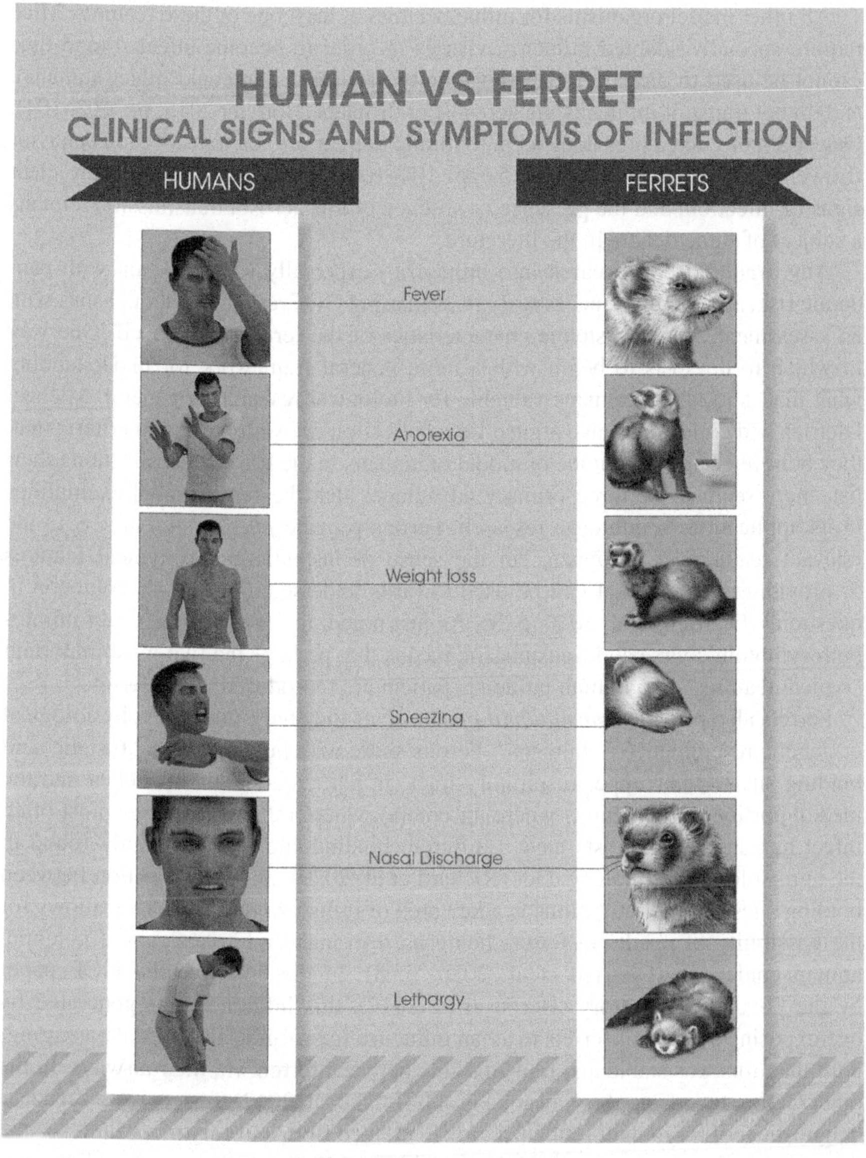

Fig. 7.2 A depiction of the symptoms of influenza in humans and ferrets. Figure 1 from Belser et al. (2016), appearing by permission from the American Society for Microbiology

lead to severe disease—and are capable of spreading the disease via both droplet and aerosol transmission—giving us a way to determine whether a strain would be highly transmissible among humans. As these are the two variables that most determine the risk of a human influenza pandemic, the ferret model has increased in importance in parallel with fears of, and research into, this risk.

All other model organisms for influenza miss at least one of these features. Mice require specially adapted influenza viruses in order to become infected (and thus cannot be used to directly test pathogenicity of viruses that could infect humans), and do not transmit influenza via aerosols or droplets (Bouvier 2015, pp. 103–104). One strain of guinea pigs does seem to mimic human aerosol transmission behavior (Lowen et al. 2006; Bouvier 2015, pp. 104–105), but do not exhibit many clear signs of infection, and the presence or absence of this aerosol transmission remains a subject of some debate in the literature.

Any evaluation of research into influenza—especially in connection with pandemic risk, and therefore necessarily in evaluating GOF research—should start with an assessment of the epistemic characteristics of the ferret model itself. One way in which to do so is to begin with a more general framework for understanding what makes model organisms valuable for biological research in general. Michael Dietrich and colleagues have argued for a collection of twenty different criteria that, they believe, shape the choice of model organisms in the life sciences. Among their list, there seem to be three primary advantages that the ferret model exemplifies in its application to influenza research. Ferrets provide *phenomenal access* to the relevant features of influenza, "in the sense of instantiating its typical features or providing insights that can be used towards understanding the phenomenon in question" (Dietrich et al. 2020, p. 8). As just noted, the fact that the ferret mimics both symptom course and transmission means that perhaps the two most important "typical features" of a human influenza patient are reproduced in the ferret.

Ferrets also provide *translational potential*, as they have relevant "physiological or genetic resemblance to humans." Ferrets share with humans the $\alpha 2,6$ sialic acid binding site in their upper respiratory tract (in particular, in their trachea and the medial portion of their lungs where air enters), where influenza viruses most often infect humans. Avian hosts have a different binding site, predominantly found in the human lower respiratory tract (Richard et al. 2020), and the transition between binding sites through mutations is a key area of influenza research. This allows for the possibility of results in ferrets being used to make inferences about potential human epidemics (Dietrich et al. 2020, p. 8). In the controversial GOF paper by the Fouchier laboratory (Herfst et al. 2012), this inference was generated by overexposing a series of ferrets to avian influenza (a process called serial passaging) until the virus was stable and transmissible between ferrets via their airways. In the other GOF article from the Kawaoka group (Imai et al. 2012), ferrets were used as a diagnostic tool for mutant influenza viruses created from a library of variants, where the success of each variant was determined in part by its ability to cause disease in ferrets, and in part by in vitro inference in human sera.

Ferrets also offer *responsiveness*, in Dietrich and colleagues' sense of "offering better opportunities for the experimental manipulation of features of interest" (Dietrich et al. 2020, p. 5)—particularly because the long respiratory tract of the ferret permits a clear distinction between infections in the upper and the lower respiratory portions, enabling scientists to differentiate between flu viruses that are of greater risk to humans in virtue of easier access to upper respiratory tract cells. Ferrets are also large enough to collect accurate clinical data such as temperature variation, another important variable in determining case outcomes for influenza infections among human patients (Maher and DeStefano 2004).

But the literature about the use of ferrets as a model for the study of influenza is remarkably self-reflective. The Influenza Division of the National Center for Immunization and Respiratory Diseases (US) regularly publishes articles describing the nature and limits of inferences from the ferret model—and those limits are significant. To stay with the Dietrich et al. criteria as a helpful schema for categorizing these worries, there are substantial, recognized issues with (at the very least) seven of the other factors they identify when practicing scientists evaluate model organisms. Ferrets often present difficulties for both *ease of supply* and *financial costs*, as initial expenses for procuring ferrets are higher, ferret breeding programs are not often present in veterinary facilities, standard equipment for the husbandry of ferrets is not widely available from suppliers, and influenza is occasionally already endemic in breeders' populations of ferrets (Radigan et al. 2015; Oh and Hurt 2016; Maher and DeStefano 2004, p. 52). Even in their early work, Smith et al. (1933, p. 66) caution that—in addition to influenza—quarantining ferrets for 14 days is necessary due to the presence of other diseases such as distemper virus. *Ethical considerations* are more significant in ferrets—not because they are necessarily taken to be a more ethically sensitive experimental subject, but rather thanks to the general lack of "best practices" for husbandry and experimentation that one might find in systems such as mice, often slowing animal ethics committee approval (Oh and Hurt 2016, p. 3).

Standardization is often lacking in ferret research, as highly inbred strains of ferrets have not been developed in the way that they have been in rodents, leading to a largely unknown amount of genetic diversity between populations (Maher and DeStefano 2004, p. 52). The *viability* and *durability* of ferrets is often limited, as their cost often means that sample sizes of as few as five animals are regularly used, even for research with potentially significant social impact (Oh and Hurt 2016, p. 3), and stocks of animals are not usually maintained by the laboratories performing these studies. Both the *availability of methods and techniques*, as well as our extant body of *epistemic resources* for ferret experiments are also underdeveloped. On the practical side, viral inoculation methods, study endpoints, and necropsy methods are not necessarily shared across all labs (Belser et al. 2020). Ferret-specific reagents are often commercially unavailable (Belser et al. 2016, p. 733). More broadly, the ferret genome was only sequenced fairly recently (Peng et al. 2014)—notably, well after the controversial gain-of-function research discussed above was announced in late 2011 (Enserink 2011).

To sum up, then, ferrets are a difficult, expensive, and minimally standardized system. Overall assessments of the effectiveness of the ferret model are therefore often quite guarded, even by virology researchers themselves. Belser et al. write that "these confounders result in heterogeneity with regard to procedures and practices established at all levels of research, from individual investigators or institutions to broad country-specific regulations" (Belser et al. 2020, p. 12). Elsewhere, they note that "with no universal standard established for ferret studies evaluating the relative risk of influenza viruses, virus-specific, host-specific, and equipment-specific variation is inherent in published studies" (Belser et al. 2016, p. 734). Oh and Hurt write that

> the use of ferrets for influenza studies has been limited by factors such as animal availability, genetic heterogeneity (out-bred), the requirement of a complex husbandry facility and caging system, and a lack of immunological reagents and genetically modified mutants for immunological investigation. [...] Ideally, a larger number of ferrets should be used but limitations such as high experimental cost, low animal availability, limited caging capacity and ethical constraint, typically restricts most studies to group sizes of five or less ferrets. (Oh and Hurt 2016, p. 3)

The ferret is therefore, at the very least, a peculiar model organism, and one which presents us a number of *prima facie* reasons for caution about the knowledge obtained in its use. While remaining the only system capable of modeling both transmission and infection behaviors in influenza research, the scientific community itself recognizes that the system has many quirks and flaws. And indeed, a contributing factor to these challenges is that while the ferret has been a staple of influenza research for ninety years, ferret investigations have really only begun to mature since 2008, with increased effort to understand emerging pandemic influenza strains such as H5N1, H7N9, and the A(H1N1)pdm09 virus that caused the 2009 "swine flu" pandemic (Oh and Hurt 2016, pp. 3–4).

Of course, this does not constitute an argument that ferrets are particularly unusual in this regard. Scientists will often have very good reasons to use what might appear from the outside to be "better" and "worse" model organisms in different contexts. As Dietrich et al. put it, "in any instance of organismal choice, not all of the criteria will be used; of those used, not all will be given equal weight. In fact, many research situations require scientists to prioritize a cluster of criteria over others" (Dietrich et al. 2020, p. 9). But, as they go on to explain, recognizing the points of conflict between these criteria reflects something important about what it means for a model organism to be a "good" one, and what kinds of goals a scientific community has targeted in their research. It is in debate over precisely these questions, I argue, that we will find appeals to integrative promise.

7.4 Experimental and Model Organisms

In particular, I want to consider here the position of ferrets with respect to a distinction sometimes drawn between "model organisms" in a robust sense and "experimental" or "Krogh organisms" (following Green et al. 2018) in a much narrower vein. By contrast with model organisms as I laid them out at the beginning of the chapter, experimental organisms "are typically not standardized; little effort is put into building specialized infrastructures around them; and they are often chosen for specialized features that allow for investigation of specific biological problems" (Green et al. 2018, p. 5; see also Ankeny and Leonelli 2020, p. 7). Green et al. baptize these "Krogh organisms" after the eponymous principle first coined by August Krogh: "for a large number of problems there will be some animal of choice, or a few such animals, on which it can be most conveniently studied" (quoted in Green et al. 2018, p. 1). These organisms are usually those where a given trait is "more accessible or prominent," often in the form either of "extreme adaptations" (i.e., unusual capabilities not often found elsewhere in the animal kingdom) or of organisms that enable "exploring variation over the identified physiological features" (i.e., unusual variability that can help us understand the landscape of biological possibility; Green et al. 2018, p. 8). As an example of the former, Green et al. discuss the kidney adaptations of desert rodents to produce concentrated urine by absorbing extra water. We try to use these extreme adaptations, then, to shed light on the function of those same parts in other organisms. For the latter, among others, they note that comparisons between unusual adaptations—like the echolocation machinery in bats and in whales—can show us vital information about the functional constraints governing the evolution of traits.

This distinction seems, *a priori*, to be particularly interesting for the case of ferrets. First, ferrets began their career as a custom-tailored, single-purpose model for exploring a very narrow set of phenomena associated with influenza symptoms and transmission. That seems precisely like an experimental organism, insofar as ferrets make a particular feature (respiratory disease transmission) "more accessible or prominent." They also *don't* seem to be a traditional model organism—they don't feature on the lists as published by organizations like the NIH, and they don't have, as we have already seen, the kind of institutional, commercial, and technical backing of organisms like the mouse. These features might lead us to think that they are experimental organisms.

On the other side, however, ferrets do not seem to offer "extreme" adaptations of the sort that are often described for Krogh organisms—as they share the same binding structures in their upper respiratory system, we do not need a special argument to ground the mere possibility of translating those results to our understanding of humans, as we might for the translation of study results from desert rodents. That might mean that the category of Krogh organisms is less useful here. Making things even more complicated, as we will see below, scientists working on ferret research often actively push for the standardization and the expansion in scope of the ferret model, in the way that we might expect from the development of a model organism rather than experimental organism. To be sure, this remains for the most part focused on respiratory viruses, with respiratory syncytial virus (RSV) and coronaviruses (including SARS-CoV-2) also studied in ferrets (Muñoz-Fontela et al. 2020), but development of the model for other viral infections (Schiffman et al. 2020) and more general immunological studies is now underway (Hundakova et al. 2022).

This seems to indicate that ferrets hold a delicate and confusing place between experimental and model organisms, a fact that might help us understand the peculiar epistemic status of the ferret as a model system. The standards for evaluating an organism as an experimental organism and those for a model organism will certainly differ—Green et al., for instance, describe the "virtues" of experimental organisms as "complimentary" to those of traditional model organisms (Green et al. 2018,

p. 19). Nor is this demarcation just a question of terminology. As Ankeny and Leonelli put it:

> Given the significance of the label, questions about whether and in what sense a particular experimental organism is a 'model organism' require explorations that go beyond abstract philosophical analyses or laboratory boundaries; answers to these questions have clear epistemic, social, political, and economic implications with regard to how science is conducted and how knowledge is constructed. (Ankeny and Leonelli 2020, p. 4)

Being marked out as a model organism comes with a healthy dose of institutional and social clout, as well as epistemic weight accorded to the results derived from such organisms. Moving ferrets up the continuum from experimental toward model organisms would thus be beneficial to ferret researchers for a variety of reasons, a fact which those researchers keenly recognize.

This process sits uneasily with the repeated recognition of the uncertainty present in the ferret literature. As has been noticed by a number of critics of model organisms, this sort of nuance is often driven out of our perspective surrounding traditional model organisms like the mouse (Ankeny 2010). The ferret thus represents a potential site of extreme interest for philosophers exploring these kinds of questions: neither clearly an experimental nor a model organism, and the target of an active and intense process of standardization (at work in different ways in different places around the globe), it is related in a variety of complex ways to a host of socially significant scientific and public-health problems.

7.5 Model Organisms and Model Inferences

How, then, do ferret researchers evaluate the kinds of inferences that they draw on the basis of their model, and how does this evaluation relate to the status of ferrets as experimental or model organisms? Strangely, given its social and ethical importance, analyses of the nature of GOF research in ferrets is often absent or underdeveloped. In the risk and benefit assessment of GOF research commissioned by the National Institutes of Health, for example, in assessing whether GOF research will answer the question "can animal influenza viruses become transmissible between humans," the report simply states that "the key limitations of this approach are that observations in animal models may not translate to humans and that the adaptive changes observed in the laboratory may not be possible in nature" (Gryphon Scientific 2016, p. 313). These kinds of anodyne translational worries are present in any animal research, and capture nothing of the complex landscape described in the previous two sections.

Even in places where some attention is paid to these dimensions of epistemic uncertainty, the character of the scientific knowledge developed in GOF contexts is often not questioned. For instance, I agree wholeheartedly with Casadevall et al. that "the debate [around GOF] has largely ignored the question of the epistemological value of such experiments" (Casadevall et al. 2014, p. 1). But when they turn to the nature of the knowledge that GOF is taken to produce, they write that GOF experiments can be defended "because they yield information that is consistent with the normative standards of the fields of microbiology and infectious diseases, and as such, they provide information that is immediately accessible and interpretable in the context of standards in the field" (Casadevall et al. 2014, p. 2).

This evaluation is, of course, correct: articles on GOF research adhere to the kinds of best practices that the field has developed as a whole; there is no reason to believe that they have not been peer-reviewed in detail (indeed, one might surmise that they would face *more* scrutiny than average); and they are highly cited and widely discussed. But noting these social and contextual facts does not amount to the kind of weighing of the "epistemological value" of experiments in ferrets for which Casadevall et al. called in the first place. Researchers working within virology itself recognize that there are some good reasons to believe that ferret research might not offer us the type of quality guarantees available in other domains in the contemporary life sciences, and the risk and benefit assessment mentioned above noted, for numerous scientific applications of GOF research, that the social and contextual facts "are constrained by scientific uncertainties associated with the data" (Gryphon Scientific 2016, pp. 251–253).

A number of paths are open to us here. We might pursue the (more purely) ethical consequences of taking on the risk entailed by work that operates within this kind of uncertainty (a project that a colleague and I have undertaken elsewhere; Evans and Pence 2024). Such considerations might tempt one to more carefully consider, or even support, non-GOF alternatives that could offer us the same kinds of benefits while radically reducing the potential costs (Lipsitch and Galvani 2014; Evans 2018). But for my purposes here, considering ferrets in the light of integrative promise, I want to turn back toward their peculiar status as experimental/model organisms and how this might relate to the uncertainty in the discussion over the interpretation of these results.

As I mentioned briefly above, the ferret literature is surprisingly self-reflective about the limits of their model. But ferret researchers don't rest content with the current state of play. Many of the articles that mention these limits make appeals precisely to the necessity of overcoming them—for, as we can now more clearly state the case, "upgrading" the ferret from an experimental organism to a fully-fledged model organism suitable for the kind of integrative work that I described in Sect. 7.1. While recognizing that a diverse collection of methodological approaches might allow for beneficial avoidance of bias or lock-in effects, Belser et al. argue nevertheless that "understanding the role that these choices in experimental design can play in research studies and how these mammalian models can be further improved for use in risk assessment is critical" (Belser et al. 2016, p. 734). They conclude their paper by writing that

> Greater precision in descriptions of experimental methods when publishing and sharing research will improve the ability to conduct larger-scale analyses of trends present among particular virus subtypes or other features with statistical rigor not generally possible in individual studies. (Belser et al. 2016, p. 741)

Put differently, if we want to guarantee the possibility that ferrets will serve as a vertically-integrative "platform" in Griesemer's sense, we need to move much more quickly through the process of standardization and abstraction described as so important by Ankeny and Leonelli. We need further investment if we are to capitalize on ferrets' integrative promise.

This recommendation was directly carried out by the group at the Centers for Disease Control and Prevention (US) that published the article cited above, along with ten other laboratories across the world. Frustrated with both their experimental limitations and their lack of understanding of the impact of those very limitations, these eleven research groups performed a large longitudinal study, published in 2022, deploying identical experiments, with identical strains of influenza (selected from previously characterized stock H1N1 viruses), in a wide-ranging and extremely impressive effort to chart the effects of the kinds of variability just discussed on the results that each laboratory would obtain (Belser et al. 2022). On the one hand, the outcomes were largely positive. All parameters that were thus far known to affect transmission were standardized, including inoculation methods, viral doses, volumes, and respiratory droplet transmission methods. A number of other environmental parameters were analyzed and monitored, including airflow, temperature, relative humidity, and distance between cages. Overall results were "concordant, suggesting the robustness of the ferret model for zoonotic influenza risk assessment" (Belser et al. 2022, p. 1). But even then, the authors suggest caution when making inferences toward socially relevant outcomes: three laboratories would need to agree to confirm high transmissibility of a new strain of influenza, and five laboratories would be required to confirm a judgment of low transmissibility. It also remains unclear whether the collection of parameters adopted in this study will become standard practice, either within the laboratories implicated in the study itself or globally. This kind of negotiation over the stringency-opportunism trade-off is expected in an environment where the integrative promise of ferrets remains uncertain. Ferret researchers are actively working to justify the status of the ferret as an integration-supporting platform, and thus to make the case stronger for the integrative promise of the ferret model.

It's important to remember that the integrative promise of ferrets has genuine, real-world consequences. Even if the epistemic credentials of ferrets aren't sufficient to guarantee the kind of horizontal integration that could make them a model organism applicable across the tree of life, both GOF experiments with which we began our discussion did, nonetheless, result in a mammalian-transmissible form of influenza, that in principle could spread in humans. So whether or not ferrets give us the right kind of information we need to know *why* pandemic influenzas arise in humans, and defend against it, we have still constructed a mammalian transmissible strain of influenza with its attendant risks. Put differently, the grand challenge that ferret work aims to solve is self-evident, but its ability to do so, and to do so with a likelihood which makes it worth taking on the dangers it entails, will be ensured by the existence of stringent tests just like those evaluated in this longitudinal study.

Other researchers are promoting the promise of the ferret for further horizontally integrative work, serving as a point of comparison for many human and animal disease conditions. Taking as their starting point not the 1933 discovery of the susceptibility of ferrets to influenza, but rather a 1911 paper using ferrets as an embryological model, they argue for the importance of ferrets in "cardiac research, spinal cord injury, and several lung conditions, including smoke-induced chronic obstructive pulmonary disease (COPD), cystic fibrosis, and tobacco-induced lung cancer" (Albrecht et al. 2018, p. 1).[4] Standardized ferret immunological reagents and some cell lines are under development, as is validation for protocols to help ensure reproducible results across laboratories.

In short, the ferret model is growing up. Situated as it is between an experimental organism—a limited-use model of a particular set of phenomena, namely the transmission of infectious diseases—and a proper model organism, its advocates (the acknowledgements section of Albrecht et al. 2018 refers to them as Team Ferret) appeal for greater investment, exactly as one would expect if these were explicit efforts to move the ferret toward a role as a "mini" or "nascent" model organism—assuredly one with a more limited scope than the mouse, but nonetheless carrying the kind of epistemic and representational flexibility characteristic of model rather than experimental organisms. That investment would make ferrets easier to use and more opportunistically desirable; deploy reproducible and standard protocols and subject those to extensive testing, giving us a better idea of those domains in which (like respiratory infection) ferrets could provide stringent tests of integrative explanations; and ensure that ferrets do indeed make good on their promise to vertically and horizontally integrate biological knowledge. It would allow us to demonstrate the very integrative promise that, ferret advocates argue, they have had all along as, to borrow an old joke from the literature, Biology's Next Top Model (Maher 2009).

As we saw above, there was something somewhat confusing about the way that scientists have, on the one hand, defended both the scientific and the social importance of GOF research while, at the same time, downplaying fears about its impact by casting doubt on whether or not the viruses thus created could actually infect human beings. This confusion becomes more explicable when viewed through the lens of the integrative promise of GOF research. As I've tried to argue throughout this work, the epistemic and social aspects of integrative promise cannot be simply dissociated from one another. The social value envisioned by the pursuit of GOF research—its ability to help us attack the grand challenge of future zoonotic pandemics—is part and parcel of the integrative promise of pursuing that research further. And this remains the case even if, with respect to any particular experiment performed today, we may not yet have the confidence that ferret model research is stringent enough to support an inference to the transmissibility to human beings of a particular virus created in a laboratory. To draw an analogy with the case that I

[4] They also note, paralleling the emphasis placed by Leonelli and Ankeny (2012) on the importance of online databases for model organism research, that an online database of ferret neural connectome data has recently been launched (Sukhinin et al. 2016), offering the possibility for future use of ferrets in neuroscience.

presented in Chap. 3, we don't yet know whether we're in the position of Darwin (successfully) studying orchids, or Darwin (unsuccessfully) studying earthworms. And it is for precisely this reason that the integrative promise of ferret model research is the subject of so much heated debate.

Model organisms, therefore, not only serve the ends of vertical and horizontal integration in contemporary life science, but moreover are—at least in cases like the ferret—promoted and critiqued in terms of their integrative promise. Considering integrative promise as a whole can help us understand the complex interplay of social and epistemic aspects in the evaluation of these research programs in virology, yet another advantage of identifying this pattern in appeals to explanatory virtues.

References

Albrecht RA, Liu WC, Sant AJ, Tompkins SM, Pekosz A, Meliopoulos V, Cherry S, Thomas PG, Schultz-Cherry S (2018) Moving forward: recent developments for the ferret biomedical research model. mBio 9(4):10.1128/mbio.e01113-18. https://doi.org/10.1128/mBio.01113-18

Ankeny RA (2010) Historiographic reflections on model organisms: or how the mureaucracy may be limiting our understanding of contemporary genetics and genomics. Hist Philos Life Sci 32(1):91–104

Ankeny RA, Leonelli S (2020) Model organisms. Cambridge University Press, Cambridge

Belser JA, Katz JM, Tumpey TM (2011) The ferret as a model organism to study influenza A virus infection. Dis Models Mech 4(5):575–579. https://doi.org/10.1242/dmm.007823

Belser JA, Eckert AM, Tumpey TM, Maines TR (2016) Complexities in ferret influenza virus pathogenesis and transmission models. Microbiol Mol Biol Rev 80(3):733–744. https://doi.org/10.1128/MMBR.00022-16

Belser JA, Eckert AM, Huynh T, Gary JM, Ritter JM, Tumpey TM, Maines TR (2020) A guide for the use of the ferret model for influenza virus infection. Am J Pathol 190(1):11–24. https://doi.org/10.1016/j.ajpath.2019.09.017

Belser JA, Lau EHY, Barclay W, Barr IG, Chen H, Fouchier RAM, Hatta M, Herfst S, Kawaoka Y, Lakdawala SS, Lee LYY, Neumann G, Peiris M, Perez DR, Russell C, Subbarao K, Sutton TC, Webby RJ, Yang H, Yen HL, Working group on the standardization of the ferret model for influenza risk assessment, Blanchard E, Maines TR, Choy KT, Sia SF, Su W, Frise R, Zhou J, Hurt AC, Li C, Bestebroer T, de Meulder D, Lexmond P, Chiba S, Bhagwat A, Jones JE, Kormuth KA, Le Sage V, Caceres CJ, Carnaccini S, Ferreri LM, Geiger G, DeBeauchamp JL, Hu M, Jeevan T, Kercher LA, Patel DR, Septer KM, Sim DG (2022) Robustness of the ferret model for influenza risk assessment studies: a cross-laboratory exercise. mBio 13(4):e01174-22. https://doi.org/10.1128/mbio.01174-22

Bouvier NM (2015) Animal models for influenza virus transmission studies: a historical perspective. Curr Opin Virol 13:101–108. https://doi.org/10.1016/j.coviro.2015.06.002

Casadevall A, Imperiale MJ (2014) Risks and benefits of gain-of-function experiments with pathogens of pandemic potential, such as influenza virus: a call for a science-based discussion. mBio 5(4):10.1128/mbio.01730-14. https://doi.org/10.1128/mBio.01730-14

Casadevall A, Howard D, Imperiale MJ (2014) An epistemological perspective on the value of gain-of-function experiments involving pathogens with pandemic potential. mBio 5(5):10.1128/mbio.01875-14. https://doi.org/10.1128/mBio.01875-14

Dietrich MR, Ankeny RA, Crowe N, Green S, Leonelli S (2020) How to choose your research organism. Stud Hist Philos Biol Biomed Sci 80:101227. https://doi.org/10.1016/j.shpsc.2019.101227

Enserink M (2011) Scientists brace for media storm around controversial flu studies. Science Insider. https://www.science.org/content/article/scientists-brace-media-storm-around-controversial-flu-studies

European Molecular Biology Laboratory (2022) Molecules to ecosystems: the EMBL Programme 2022–2026. Tech. rep. EMBL, Heidelberg. https://www.embl.org/about/programme/

Evans NG (2013) Great expectations—ethics, avian flu and the value of progress. J Med Ethics 39(4):209–213. https://doi.org/10.1136/medethics-2012-100712

Evans NG (2018) Ethical and philosophical considerations for gain-of-function policy: the importance of alternate experiments. Front Bioeng Biotechnol 6. https://doi.org/10.3389/fbioe.2018.00011

Evans NG, Pence CH (2024) Gain-of-function research and model organisms in biology. J Med Ethics 50(3):201–206. https://doi.org/10.1136/jme-2022-108853

Evans NG, Lipsitch M, Levinson M (2015) The ethics of biosafety considerations in gain-of-function research resulting in the creation of potential pandemic pathogens. J Med Ethics 41(11):901–908. https://doi.org/10.1136/medethics-2014-102619

Evans NG, Selgelid MJ, Simpson RM (2022) Reconciling regulation with scientific autonomy in dual-use research. J Med Philos 47(1):72–94. https://doi.org/10.1093/jmp/jhab041

Green S, Dietrich MR, Leonelli S, Ankeny RA (2018) 'Extreme' organisms and the problem of generalization: interpreting the Krogh principle. Hist Philos Life Sci 40(4):65. https://doi.org/10.1007/s40656-018-0231-0

Griesemer J (2013) Integration of approaches in David Wake's model-taxon research platform for evolutionary morphology. Stud Hist Philos Biol Biomed Sci 44(4):525–536. https://doi.org/10.1016/j.shpsc.2013.03.021

Gryphon Scientific (2016) Risk and benefit analysis of gain of function research. Tech. rep. US National Institutes of Health. https://www.gryphonscientific.com/resources/gain-of-function/

Herfst S, Schrauwen EJA, Linster M, Chutinimitkul S, de Wit E, Munster VJ, Sorrell EM, Bestebroer TM, Burke DF, Smith DJ, Rimmelzwaan GF, Osterhaus ADME, Fouchier RAM (2012) Airborne transmission of influenza A/H5N1 virus between ferrets. Science 336(6088):1534–1541. https://doi.org/10.1126/science.1213362

Hundakova A, Leva L, Toman M, Knotek Z (2022) A ferret model of immunosuppression induced with dexamethasone. Vet Immunol Immunopathol 243:110362. https://doi.org/10.1016/j.vetimm.2021.110362

Imai M, Watanabe T, Hatta M, Das SC, Ozawa M, Shinya K, Zhong G, Hanson A, Katsura H, Watanabe S, Li C, Kawakami E, Yamada S, Kiso M, Suzuki Y, Maher EA, Neumann G, Kawaoka Y (2012) Experimental adaptation of an influenza H5 HA confers respiratory droplet transmission to a reassortant H5 HA/H1N1 virus in ferrets. Nature 486(7403):420–428. https://doi.org/10.1038/nature10831

Kilianski A, Nuzzo JB, Modjarrad K (2016) Gain-of-function research and the relevance to clinical practice. J Infect Dis 213(9):1364–1369. https://doi.org/10.1093/infdis/jiv473

Leonelli S (2016) Data-centric biology: a philosophical study. University of Chicago Press, Chicago

Leonelli S, Ankeny RA (2012) Re-thinking organisms: the impact of databases on model organism biology. Stud Hist Philos Biol Biomed Sci 43(1):29–36. https://doi.org/10.1016/j.shpsc.2011.10.003

Levy A, Currie A (2015) Model organisms are not (theoretical) models. Br J Philos Sci 66(2):327–348. https://doi.org/10.1093/bjps/axt055

Li FCK, Choi BCK, Sly T, Pak AWP (2008) Finding the real case-fatality rate of H5N1 avian influenza. J Epidemiol Community Health 62(6):555–559. https://doi.org/10.1136/jech.2007.064030

Lipsitch M (2018) Why do exceptionally dangerous gain-of-function experiments in influenza? In: Yamauchi Y (ed) Influenza virus: methods and protocols, methods in molecular biology. Springer, New York, pp 589–608. https://doi.org/10.1007/978-1-4939-8678-1_29

Lipsitch M, Galvani AP (2014) Ethical alternatives to experiments with novel potential pandemic pathogens. PLoS Med 11(5):e1001646. https://doi.org/10.1371/journal.pmed.1001646

Lowen AC, Mubareka S, Tumpey TM, Garcia-Sastre A, Palese P (2006) The guinea pig as a transmission model for human influenza viruses. Proc Natl Acad Sci 103(26):9988–9992. https://doi.org/10.1073/pnas.0604157103

Maher B (2009) Evolution: biology's next top model? Nature 458(7239):695–698. https://doi.org/10.1038/458695a

Maher JA, DeStefano J (2004) The ferret: an animal model to study influenza virus. Lab Animal 33(9):50–53. https://doi.org/10.1038/laban1004-50

Maudlin T (2007) The metaphysics within physics. Oxford University Press, Oxford

Muñoz-Fontela C, Dowling WE, Funnell SGP, Gsell PS, Riveros-Balta AX, Albrecht RA, Andersen H, Baric RS, Carroll MW, Cavaleri M, Qin C, Crozier I, Dallmeier K, de Waal L, de Wit E, Delang L, Dohm E, Duprex WP, Falzarano D, Finch CL, Frieman MB, Graham BS, Gralinski LE, Guilfoyle K, Haagmans BL, Hamilton GA, Hartman AL, Herfst S, Kaptein SJF, Klimstra WB, Knezevic I, Krause PR, Kuhn JH, Le Grand R, Lewis MG, Liu WC, Maisonnasse P, McElroy AK, Munster V, Oreshkova N, Rasmussen AL, Rocha-Pereira J, Rockx B, Rodríguez E, Rogers TF, Salguero FJ, Schotsaert M, Stittelaar KJ, Thibaut HJ, Tseng CT, Vergara-Alert J, Beer M, Brasel T, Chan JFW, García-Sastre A, Neyts J, Perlman S, Reed DS, Richt JA, Roy CJ, Segalés J, Vasan SS, Henao-Restrepo AM, Barouch DH (2020) Animal models for COVID-19. Nature 586(7830):509–515. https://doi.org/10.1038/s41586-020-2787-6

Oh DY, Hurt AC (2016) Using the ferret as an animal model for investigating influenza antiviral effectiveness. Front Microbiol 7. https://doi.org/10.3389/fmicb.2016.00080

O'Malley MA, Soyer OS (2012) The roles of integration in molecular systems biology. Stud Hist Philos Biol Biomed Sci 43(1):58–68. https://doi.org/10.1016/j.shpsc.2011.10.006

Peng X, Alföldi J, Gori K, Eisfeld AJ, Tyler SR, Tisoncik-Go J, Brawand D, Law GL, Skunca N, Hatta M, Gasper DJ, Kelly SM, Chang J, Thomas MJ, Johnson J, Berlin AM, Lara M, Russell P, Swofford R, Turner-Maier J, Young S, Hourlier T, Aken B, Searle S, Sun X, Yi Y, Suresh M, Tumpey TM, Siepel A, Wisely SM, Dessimoz C, Kawaoka Y, Birren BW, Lindblad-Toh K, Di Palma F, Engelhardt JF, Palermo RE, Katze MG (2014) The draft genome sequence of the ferret (Mustela putorius furo) facilitates study of human respiratory disease. Nat Biotechnol 32(12):1250–1255. https://doi.org/10.1038/nbt.3079

Radigan K, Budinger S, Misharin A, Chi M (2015) Modeling human influenza infection in the laboratory. Infect Drug Resist 8:311–320. https://doi.org/10.2147/IDR.S58551

Richard M, van den Brand JMA, Bestebroer TM, Lexmond P, de Meulder D, Fouchier RAM, Lowen AC, Herfst S (2020) Influenza A viruses are transmitted via the air from the nasal respiratory epithelium of ferrets. Nat Commun 11(1):766. https://doi.org/10.1038/s41467-020-14626-0

Schiffman Z, Liu G, Cao W, Zhu W, Emeterio K, Qiu X, Banadyga L (2020) The ferret as a model for filovirus pathogenesis and countermeasure evaluation. ILAR J 61(1):62–71. https://doi.org/10.1093/ilar/ilab011

Schultz-Cherry S, Webby RJ, Webster RG, Kelso A, Barr IG, McCauley JW, Daniels RS, Wang D, Shu Y, Nobusawa E, Itamura S, Tashiro M, Harada Y, Watanabe S, Odagiri T, Ye Z, Grohmann G, Harvey R, Engelhardt O, Smith D, Hamilton K, Claes F, Dauphin G (2014) Influenza gain-of-function experiments: their role in vaccine virus recommendation and pandemic preparedness. mBio 5(6):10.1128/mbio.e02430-14. https://doi.org/10.1128/mBio.02430-14

Smith W, Andrewes CH, Laidlaw PP (1933) A virus obtained from influenza patients. Lancet 222(5732):66–68. https://doi.org/10.1016/S0140-6736(00)78541-2

Sukhinin DI, Engel AK, Manger P, Hilgetag CC (2016) Building the Ferretome. Front Neuroinform 10. https://doi.org/10.3389/fninf.2016.00016

Chapter 8
Using Papers as Data

Textual Analysis and Tools for Integrative Promise

Abstract In this chapter, I use the lens of integrative promise to offer an explanation of a fact that is often noted in passing but rarely considered explicitly in the literature: while, in recent years, essentially every scientific discipline has turned to textual analysis of the journal literature as a tool, such uses are far more prominent in the life sciences than in other scientific fields. Why might this be? I argue here that this results from the way in which textual analysis of the scientific literature can contribute to the discovery of promising explanations in biology. Such a connection is not necessary—textual analysis can certainly be used for other purposes as well—but the importance of integrative promise in the life sciences can thus help us understand why textual analysis would have so quickly become such an important tool in this context.

Keywords Textual analysis · Digital humanities · Literature-based discovery · Opportunism · Integration · Natural language processing · Data mining

In almost all of the previous examples thus far in the book, I've tried to demonstrate how the work of practicing scientists makes direct reference to integrative promise. Put differently, I've tried to show that explanations across the life sciences do indeed instantiate this linked pattern of explanatory virtues, and that doing so is a crucial part of the defense of those explanations. I've also argued that using this frame to look at some well-known, particular episodes in biology, like the debate over natural history, or the use of model organisms or big data in biology, offers some advantages and novel perspectives compared to our current approach.

In this chapter, I want to take a slightly different angle. What I want to do now is pose a question about a practice that, to my knowledge, hasn't yet been directly addressed in the literature, concerning the use of digital humanities and textual analysis in the life sciences: why is it that these methods are used more often in the life sciences than in other disciplines? I then want to propose that, even though there is no necessary connection between textual analysis and integrative promise, and promise is not directly discussed in defending the use of such tools, a compelling answer to this question—and a clarifying understanding of what textual analysis is

C. H. Pence, *Integrative Promise*, Synthese Library 499,
https://doi.org/10.1007/978-3-031-88471-9_8

doing for biologists—can be offered by exploring the role that these methods play as tools that can help us in the pursuit of promising explanations.

In other work, I've already spent much time thinking about the potential use of methods drawn from textual analysis in the philosophy of biology (Pence and Ramsey 2018; Lean et al. 2021; Pence 2022a,b). Of course, the deployment of such tools is by no means limited to philosophy. (Far from it, in fact; philosophy is one of the fields in which the uptake of digital approaches has been the slowest and has met the most opposition.) The sciences have seen a rapid expansion of such work, which takes the scientific literature itself as a source of scientific data. On the one hand, this is part and parcel of the increasing appeals to big data in science that I discussed in Chap. 6.[1] But to leave the story there misses a fact that, I think, deserves an explanation. If we set aside the social and political sciences—in which, since social media and digital communication are themselves the objects of extensive study, textual analysis has become more or less unavoidable—the domain of the sciences in which these tools are most often invoked is, by far, the life sciences.

What could explain this? To be sure, an element of any such explanation will be contingent and historical; the development of the first databases, like MEDLINE, that rendered such work possible, along with a number of early successes using them (some of which I will discuss in more detail below) assuredly created a kind of "founder effect" that made appeals to digital tools more likely. But I think that this is only part of the story. I want to argue here that seeing textual analysis as a methodology that can facilitate our search for integrative promise gives us an interesting and powerful way to think about why biologists might have been so attracted to its use. Even though literature describing this approach does not usually (or, at least, not with the same frequency as the other examples thus far in the volume have demonstrated) make direct reference to integrative promise, seeing promise as among the underlying driving forces behind the adoption of textual analysis can help us understand it better, give us guidance for normatively evaluating how it is put into practice, and offer suggestions for its improvement.

I'll first start by substantiating a few of the empirical claims that I made above: namely, that the life sciences are indeed among the scientific disciplines where textual analysis is most frequently employed, and that promise (or its components) are not often found among the justifications for its use in the literature (which, as we will see, is much more pragmatic). Then I'll turn to a detailed assessment of a few case studies of biological textual analysis, drawn especially from work on "literature-based discovery" of candidate drugs. Finally, I'll conclude by considering how integrative promise could play a helpful role in this story, focusing on how we might construe textual analysis, in at least some cases, as a key to unlocking promising explanations.

---

[1] This comparison between big data in other parts of science and textual analysis has not been lost on practitioners. David Searls, for instance, has argued that there is irony in the fact that revolutionary big-data biologists "are looking for help … in the trusty old scientific literature, and in how they are revisiting this time-honored medium, with the same sort of mindset that fomented the genomic and post-genomic revolutions in the first place" (Searls 2001, p. 88).

8.1 Textual Analysis in Biology

When compared to other sciences, the extent to which textual analysis has been adopted as an important element of biomedical research is relatively striking. Following on the methodology developed by Farrell et al. (2022), we can explore the number of papers in the Scopus database (Boyle and Sherman 2006; Mongeon and Paul-Hus 2016) that mention textual analysis or natural language processing (NLP) in their title, abstract, or article keywords.[2] Of course, this will be a strict underestimate of the number of papers that actually use such tools, as they could have been employed without being taken to be important enough to appear in these parts of the document metadata. But it should suffice for establishing the kind of general order-of-magnitude claim that I want to defend here.

What we find when we do so is that after starting to be used in the 1990s, textual analysis has exploded in the life sciences, appearing in over three thousand articles per year as of 2022 (Fig. 8.1). Its use has persistently remained around twice as prevalent as that in the physical sciences, and the idea that "the life sciences is the most common domain in text-mining research" (Przybyła et al. 2016, p. 25) is now a commonplace in the literature.

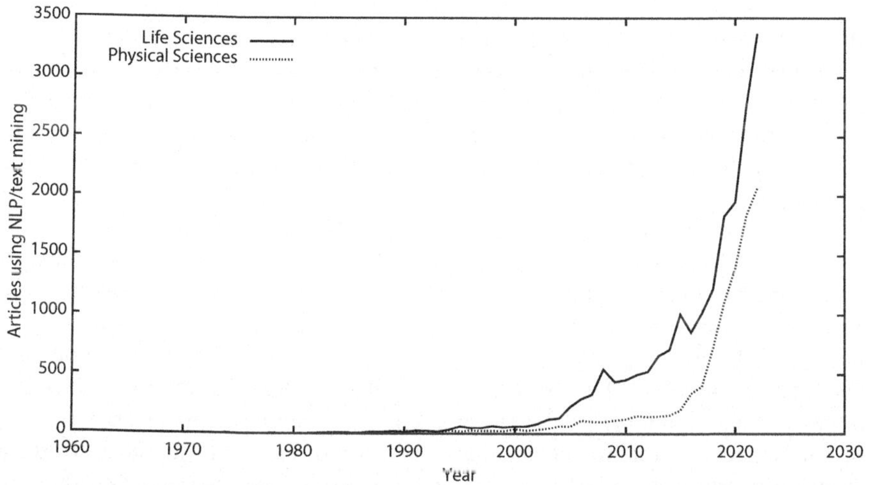

Fig. 8.1 A graph of the number of articles indexed in Scopus in the life and physical sciences making use of textual analysis, from 1965 to 2022. (Data current as of April 27, 2023.) Methodology for the creation of this graph and the raw data that generated it are available at https://doi.org/10.6084/m9.figshare.28060508

[2] More precisely, we search title, abstract, and author-provided keywords for "NLP," "natural language processing," or "text mining."

As I've briefly argued above, this strikes me as a fact about biological practice that deserves discussion. The difference between text mining's uptake in physical and life sciences doesn't seem to be directly explained by the use of big data in scientific practice: while one might see the proliferation of text analysis in biology as roughly contemporaneous with the advent of the Human Genome Project (which ran from 1990 until 2003), similar changes don't seem to have taken place around the creation of big data projects in physics, astronomy, and so forth (which, after all, date back to about the same time, with the initial development of the World Wide Web as a data-sharing tool at CERN in the late 1980s), casting doubt on any simplistic connection between big-data science and textual analysis.

The rationales that are offered by biologists when they defend the use of these methods also don't give us any way to explain their relative uptake in biological and physical sciences. Nearly every article that uses textual analysis introduces its motivation by arguing something similar to Farrell et al., who write that

> The volume of scientific literature is growing exponentially, with over three million peer-reviewed academic articles published each year. [. . .] Reading this amount of material is an insurmountable task. . . . As bodies of literature continue to grow, highly cited papers are more likely to be cited compared to recent work, which can result in slowing of scientific progress as transformative ideas are less likely to permeate and make substantive impact. Adopting computational approaches for analysis of scientific texts allows researchers to rapidly and systematically identify relevant publications. . . . (Farrell et al. 2022, p. 1)

In short, there's no more profound explanation to be offered of the adoption of textual analysis than the basic empirical observation that there is now more published literature than any single investigator can read.[3] While this is undoubtedly an important incentive, it is not one that would particularly distinguish the life sciences. Publication rates in any field now handily outstrip the (life)time of any contemporary scholar.

One feature that does seem at least somewhat specific to the life sciences, which we saw in Chaps. 6 and 7, is the extensive development of biological databases containing machine-readable concepts of diseases, genes, proteins, and so forth. Those databases can only be as reliable and complete as the information that's been loaded into them, often requiring painstaking manual work, validation, and consensus building. To be sure, the traditional biological literature also includes a vast amount of information about precisely these notions—the characterization of various genes, proteins, biochemical interactions, and so forth. That creates a straightforward need to build mechanisms (preferably automated, given the size of the task) to supplement those databases with information that is currently only available as unstructured text. As was already noted, as early as the late 2000s, in a prospective article about the future of biological text mining,

> There is a growing demand to 'translate' information from text into more computable forms and to cross-link the information with relevant biological databases. This linkage has the

[3] Readers who doubt the extent of this attitude can consult, for instance, any of Przybyła et al. (2016), Hunter (2017), Henry and McInnes (2017), Choi et al. (2018), and Alves et al. (2020).

> potential to improve the connection between the annotations in biological databases and the supporting evidence contained in the literature. (Altman et al. 2008, p. S7.1)

Textual analysis in biology can thus play a vital role in expanding and interlinking these extant biological databases, particularly in cases where the literature that might be relevant to those connections is too large to be processed manually.[4]

But the advancement of textual analysis in the life sciences doesn't seem to be simply tied to a putative mission to "complete" such databases. If anything, as we will see in the examples in the following section, the relationship between the two appears to be much more reciprocal than this: while it is certainly the case that textual analysis is sometimes used to add content to databases, it is equally often true that databases (of species names, genes, chemical compounds, diseases, and so forth) are used to tell us which features we should search for and analyze in otherwise unstructured text.

Perhaps, then, a kind of "virtuous circle" feedback loop between databases and textual analysis explains the latter's growth, where textual analysis is used to expand databases, which in turn enable more extensive use of textual analysis. This may be so, but it's still unclear why this would differentiate the life sciences from the physical sciences. If anything, contemporary physics—which often involves concepts distinguished by much more domain-specific vocabulary—might have less need of such databases, and hence should have had *lower* barriers of entry for textual analysis.[5]

In short, while we have a handful of stated motivations for using textual analysis in the life sciences, none of them seems to give us a sufficient explanation both for the way that textual analysis developed within the life sciences over the last thirty years and for why this work would be so much more common in life sciences than in physical sciences. To start to unpack these questions in more detail, we should zoom in on a particular collection of case studies.

8.2 Literature-Based Discovery

The idea of using digital analysis of the scientific literature to discover novel candidates for drugs is a surprisingly old one. The MEDLINE system from the National Library of Medicine, which contains bibliographic information for articles published in biomedical research, dates in its earliest incarnation to the late 1960s. It became easily searchable with the development of nascent telecommunications over

[4] Sometimes the curation of these links is nonetheless the product of old-fashioned close reading, as in the case of the MecCog system which infers disease mechanisms from, among other sources, manually annotated literature (Kundu et al. 2021).

[5] At least one article has made a link between the higher degree of formalization in background theories in physics and a more limited use there of ontologies for formally representing knowledge claims (Soldatova and King 2006); I lack the expertise to evaluate that argument here.

the course of the 1970s. Already by 1986, Don Swanson—a librarian working at the University of Chicago with no particular training in biomedical research—used the data available in MEDLINE to discover, first, an unexpected connection between Raynaud's syndrome and fish oil (Swanson 1986), and two years later described potential links between magnesium deficiency and migraine (Swanson 1988). Both observations would go on to form the basis of proposed therapies. He termed this kind of information "undiscovered public knowledge," which, while in some sense being "known" to the scientific community, was locked away in effectively siloed parts of the literature that he called "noncommunicating" or "noninteractive." Following Swanson's example, this gave rise to a research program now generally known as "literature-based discovery."

More precisely, the approach that Swanson pioneered is currently often described as the "ABC" method. Schematically, knowledge in drug development often takes the form of a tripartite connection between three entities. We want to learn about a compound A, which is capable of somehow modulating (causing, inhibiting, enhancing, etc.) a biochemical or genetic process B. That process B, in turn, is taken to be implicated in a disease C that we would like to treat. At least potentially, then, A could serve as a treatment for C. In his original work on migraine, for instance, Swanson (1988) noted that a number of articles had linked magnesium (A) with spreading cortical depression (B), a neurological phenomenon in which a physical "wave" of deactivated neurons moves slowly through the brain. Other papers had previously considered whether this spreading depression could be responsible for migraine symptoms (C). But surprisingly enough, it seemed as though no papers had ever explicitly made the connection between magnesium and migraine directly ($A \rightarrow C$).

We thus have, in Swanson's terms, two logically connected literatures on magnesium's role in spreading cortical depression and on the latter's role in migraine—in the sense that they are describing different aspects of what may well be the same phenomenon—that are noninteractive. While, of course, "literature mining methods only assisted with hypothesis generation or hypothesis support" (Baker and Hemminger 2010, p. 510) and thus would still need to be experimentally confirmed, when applied elsewhere, they proved to be surprisingly effective. Over the following decade, Swanson's ABC approach was built out into several specific software packages for performing this sort of search and discovery, and has particularly proven itself useful in recent years, as data about gene interaction and regulation in the scientific literature have now provided additional terms that can potentially be extracted from article abstracts and full-text and used to discover these kinds of links (see discussion in Baker and Hemminger 2010, p. 511).

Let's see how an example of this research looks today. The gene p53 is a tumor suppressor, involved in a whole host of cellular pathways that protect the body from cancer in response to stress (Soussi 2014). It is one of the most highly studied genes in cancer research, with some ninety thousand published articles found in PubMed as of 2018. A particular target of research on p53 is its interaction with kinases, "a family of evolutionarily related proteins that regulate other proteins through phosphorylation and that are an important source of drug targets" (Choi et al. 2018, pp. 10666–10667). Finding novel kinases that can affect the behavior of p53, then, is a potentially fruitful way in which to search for new cancer therapies. As Choi et al. argue, however (echoing the discussion of motivation in the last section), "the size of this body of literature defies human appraisal and thus limits the scope of current scientific hypotheses" (2018, p. 10667). They demonstrate that literature-based discovery offers us a practical way to sort through this work: by analyzing abstracts and performing only basic, inexpensive immunological screening, they found six candidate kinases, of which one was selected for further, more detailed characterization. In the end, the protein NEK2 was shown to phosphorylate and inhibit the function of p53.

Their methodology—significantly more complex than Swanson's original ABC approach—operates as follows. First, they searched the entirety of the abstracts available on PubMed (around twenty-one million papers) for any names and synonyms of known kinases, explicitly *excluding* those abstracts that already made a direct reference to p53 (so as to see how these kinases were characterized independently of their role in interacting with p53). This yielded some 240,000 papers that mentioned at least one known kinase; these were then cleaned by removing abstracts that discussed multiple kinases, removing rarely-discussed kinases, and randomly sampling among the abstracts for repeatedly well-discussed kinases. In total, papers describing 259 different kinases remained after this cleaning step, of which 23 were already known to interact with p53 (Choi et al. 2018, p. S2).

Each abstract was then turned into what digital humanists call a "bag of words": a representation that preserves just the frequency of each term that the abstract contained, discarding any information about the order of those terms or their sentence structure. These, in turn, were converted into word vectors. Think of each abstract as forming a vector in a 20,000-dimensional space, where the twenty thousand dimensions are the most common twenty thousand words or two-word pairs (called *bigrams*) in all of the abstracts. The magnitude of each vector along each dimension is simply the number of times that word or bigram appears within the abstract. These vectors can then be compared mathematically—in this case, the authors used the cosine distance between them to produce a similarity value between every pair of abstracts. Finally, we can compute not just the similarities between abstracts, but the similarities between kinases themselves, by creating vectors that are the average of every document describing that particular kinase (Choi et al. 2018, p. S3). After setting a threshold of minimum similarity for drawing an edge between two kinases, this information can be represented as a kinase similarity network (right-hand side of Fig. 8.2).

For less technically inclined readers, let me restate that a bit more colloquially. After assembling a collection of abstracts about each of some 259 human kinases, those abstracts were converted into a mathematical representation of the terms that each abstract uses. Those representations are then compared for similarity. Kinases often described using the same kinds of terms will be seen as closer together; those described using radically different terminology will be ranked farther apart. This is then converted into a network: if two kinases are often described similarly, there will be a "link" drawn between the two.

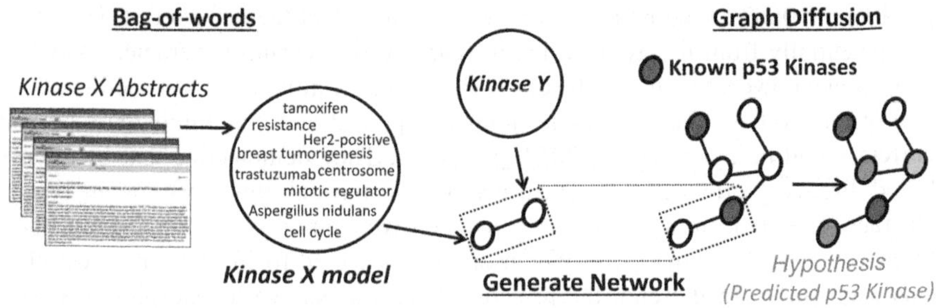

Fig. 8.2 A schematic depiction of the methodology used by Choi et al. (2018) to discover the novel inhibitory p53 kinase NEK2. Figure 1 from Choi et al. (2018, p. 10667), CC BY-NC-ND 4.0, with permission of the authors

How do we use this network to discover new potentially relevant p53 kinases? Recall that, even though abstracts that directly mentioned p53 were removed, 23 of the kinases in the network were still known to interact with p53 to at least some extent. These nodes in the network were "labeled" as being kinases of interest. Those labels are then allowed to "diffuse" through the network. Think, informally, of the nodes being marked with a dye that is allowed to "seep" out to any other kinases connected by similarity links. After a period of time, this gives us a numerical value for the predicted interaction—again, just based on the content of the literature—between that kinase and p53. Kinases linked to many known p53-interacting kinases (i.e., those whose abstracts are very similar to abstracts describing known-interacting proteins) will be ranked highest.

This inference then needs to be confirmed by biochemical analysis. Choi et al. took twenty-six of the best- and worst-ranked kinases and performed basic screening against p53, finding that binding activity of kinases with p53 was strongly correlated with the quantified prediction extracted from the literature. After choosing one of those proteins, NEK2, to study in further detail, its characterization resulted in a strong confirmation that a new p53-inhibiting kinase had been discovered. As they write in summing up their results, "this study validates a systematic approach to analyze and reason from unstructured data from the scientific literature to obtain useful scientific hypotheses" (Choi et al. 2018, p. 10670).

It would be difficult to find a more comprehensive demonstration of the potential utility of literature-based discovery (except, perhaps, the development of an NEK2-based therapy). Again, as Swanson might have outlined the case, the knowledge that NEK2 would inhibit p53 was already present in the literature, but it had never been expressed in a form that could permit the testing of that hypothesis. Once the network representation of kinase similarity made that claim accessible, it could be tested, and that testing handily confirmed the plausibility of the hypothesis.

This first example demonstrates, I think, the potential of using textual analysis in contexts where there is a massive literature covering an object of study (a dataset of 240,000 articles is quite sizable). But the connections need not necessarily be backed up by a literature that is this extensive. Literature-based discovery has also been taken up in areas where, for various practical reasons, significant empirical work is not likely to have been pursued, and we thus have a strong incentive to turn to other tools for knowledge discovery.

Consider, for instance, the case of rare diseases. By contrast with *neglected diseases*, such as malaria, which affect large numbers of individuals but are not economically profitable enough to merit the attention of major pharmaceutical companies, *rare diseases*—the prevalence of which is, on most standard definitions, well under 1 in every 10,000 individuals, sometimes even occurring only in a single known patient—"are uncommon everywhere and represent a substantial burden on individuals, families, and whole economies. As such, rare diseases demand disruptive and revolutionary drug discovery paradigms to promote the development of innovative therapies" (Alves et al. 2022, p. 491). Since it will almost never make economic sense to pursue novel drug development for a rare disease, attention in these cases has increasingly turned toward the *repurposing* of drugs. Repurposing is advantageous given the expense and, increasingly, the high rate of failure associated with novel drug development (Cuatrecasas 2006).

Even with the moderate cost of repurposing research, however, it is still unclear whether attempts to pursue it in the context of rare diseases will attract sufficient attention and funding. Literature-based discovery, then, has been proposed as a first screening step in the search for potentially underdeveloped repurposing opportunities. Initially, textual analysis can be used to quantify the extent of repurposing already present in the literature. Building on a prior database of chemical compound and disease mentions in the biomedical literature (Baker and Hemminger 2010), recent work by Baker et al. (2018) constructed a database containing around thirty-five thousand distinct chemicals and over four thousand diseases. Of these compounds, more than 60% are associated with more than one disease at different points across the literature, suggesting that these compounds were pursued for potential repurposing.[6]

As a proof of concept, Alves et al. (2022) noted that some 1267 of the 7000 drugs described in the dataset from Baker et al. (2018) were associated with at least one rare disease (Alves et al. 2020, p. 85). In a more sophisticated study, they combined the analyses of compounds from this text-mining dataset with a separate database of rare diseases and their etiology (Rappaport et al. 2017), and a more detailed classification of drug compounds as approved, investigational (in human clinical trials), and experimental (in pre-clinical or animal trials; Wishart et al. 2008, 2018). This method resulted in 754 approved drugs that could be associated with rare diseases (Alves et al. 2022, p. 494). While their methodology is presented only as a potential path forward for rare disease research, they nevertheless conclude the

[6] Notably, these kinds of analyses can't directly tell us whether or not repurposing was successful, because successful performance of a drug on the open market is almost never reflected in journal articles. The presence of clinical trials, though, can be easily detected, and one can assume that frequent discussion in the literature of a connection between a compound and a disease implies, at least weakly, a successful treatment.

paper on an optimistic note:

> We hope that the ongoing process of accumulating clinical and biomedical data and translating data to knowledge and knowledge to action in the form of data-driven and testable drug discovery hypotheses will enable the expedited and inexpensive development of new therapeutics for rare diseases in the near future. This challenge should not be left to the next generation of scientists. There is a huge opportunity to move the field and make scientific discoveries that could have a remarkable impact on the lives of people with rare diseases. (Alves et al. 2022, p. 499)

It's also noteworthy that even when relying on a number of highly automated tools, these repurposing endeavors still require that researchers are able to combine a variety of disciplinary perspectives. As Alves et al. survey the landscape,

> Computational approaches for drug repurposing include genetic associations, pathway mapping, retrospective clinical analysis, molecular docking, virtual screening, signature patching, and LBD [literature-based discovery]. Each of these approaches benefits from large-scale databases made publicly available in recent years. (Alves et al. 2022, p. 493)

Making these analyses "computational" by no means exempts researchers from the extensive background in bioinformatics, cheminformatics, and so on that is required to effectively understand the results of digital analysis.

What can we learn from the example of literature-based discovery and its use in the repurposing of drugs for the treatment of rare diseases? First—and unlike most of the other cases that I've considered in the book thus far—it's worth underlining that we're not dealing with a complete, free-standing explanation for the relevant phenomena here. Successful literature-based discovery is, at best, a first step on the road to revealing a novel therapy and its role in the mechanism that gives rise to a disease. As in cases like the connection between migraine and magnesium, it might offer us nothing more than a promising idea, an unexpected perspective that's been so-far neglected by scientists.[7]

In that sense, textual analysis stands in a different relationship to integrative promise than the other examples that I've presented earlier. Even if we might interpret the results found using these tools as proposing explanations for biological phenomena (itself dubious, as it is doubtful that the single link proposed by these methods constitutes a complete explanation), these are proposals that no biologists would accept as complete—they would never be supported without the addition of a significant quantity of further experimental data. Put differently, it seems like it would be a mistake to directly look at the explanatory virtues of the results proposed by textual analysis. What we have instead is a tool, one which will feed into the process of developing and evaluating other kinds of biological explanations. In weighing the relationship between literature-based discovery and those derived explanations, we will thus need a different kind of approach, one focused on the ways in which our tools might contribute to securing the explanatory virtues that we value.

---

[7] One is reminded of Hasok Chang's proposal of "complementary science," which encourages the contribution of scientists, philosophers, and historians to the proliferation of multiple simultaneous lines of inquiry and the pursuit of previously neglected theories and approaches (Chang 2012).

Without getting too far into the extensive literature on the role of tools in science, which would occupy more space than I can dedicate to it here, we can consider a few crucial points from the now classic analysis of "paper tools" in chemistry developed by Ursula Klein (2001). The characteristics of tools that she develops—that they are not themselves considered as direct objects of research, but as functional objects that embody accepted theory; that they are "material devices in the broadest sense of being exterior to mental processes, visible, and manoeuvrable" that "shape the kind of material traces or data gained from the target" (Klein 2001, p. 293); and that they ought to be assessed in terms of the goals of their users, as well as their capacity to generate novel goals—all apply, more or less specifically, to the use of textual analysis in biology. For biologists working on rare diseases, these kinds of discovery systems are not the direct object of study (even though they might serve as such for other researchers in information systems or natural language processing). While they are not tangible material objects, they consist of interlinked networks of websites, databases, and software packages that are largely used in much the same manner as any other physical tools with which scientists need to interface. An analysis in terms of the goals of the users of these systems is exactly the task to which I will turn momentarily. And even the brief examples that we have seen have shown us their ability to generate novel goals (e.g., to support work on drug repurposing rather than on new drug development, and to alter the operative economic incentive structure).

Let's turn then to the kind of analysis that these characteristics should drive us toward. Klein argues that understanding why authors chose a particular tool (e.g., textual analysis), why they changed it in the ways that they did (e.g., the transition from Swanson's ABC method to the more sophisticated approach of Choi et al. 2018), and how its importance developed over time (e.g., the advance of textual analysis since the 1990s and our comparison with physical sciences) requires our "discussing the collectively shared beliefs and goals of the historical actors" and "analysing how those beliefs and goals came to working life in an historically situated practice" (Klein 2001, p. 291). In short, following the interpretation for which I'll argue in the rest of the chapter, we need to think of these tools in dialogue with the pursuit of explanatory virtues.

8.3 Coming Back Around to Promise

How, then, should we see the use of textual analysis in biology as linked to the pursuit of integrative promise? First, let's very briefly run over our case studies, to see how they are related to each of the explanatory virtues that compose promise. A particularly broad sort of interdisciplinary vertical integration is assuredly at work in this research. As noted above, to make this integration possible in a semi-automated context, the process of literature-based discovery relies upon biological databases. Here, Sabina Leonelli's concept of data journeys comes into even clearer focus, as we see data from clinical studies, regulatory and clinical trial processes, human genetics, pharmacology, biochemistry, and, crucially, the corpus of scientific literature as represented by PubMed, all traveling far beyond the contexts in which they were developed for particular laboratories.

The opportunism of such methods is particularly evident, relying as they do on inferences from observations already published in the scientific literature; but equally so is the sense in which that opportunism is socially and ethically motivated—the articles that introduce these findings all do so while speaking about the importance of a "cost-effective, innovative drug discovery pipeline" that can "make drugs available more broadly and cost-effectively" (Alves et al. 2022, pp. 491, 499). To be sure, not every use of literature-based discovery invokes cases that, like these, can clearly be defined in terms of grand challenges. Appeals to social or ethical values for uses of textual analysis, however, are likely to be at least somewhat more probable than in other domains of the life sciences, given that one might not feel the need to introduce and defend unusual methodologies if a socially or ethically important problem already responded to traditional, well-financed biological methods.

The need for stringency is not explicitly discussed in this literature. As already noted, no biologist would propose the immediate green-lighting of a medication found by literature-based discovery without further experimental and clinical testing, precisely because these detections of signal in the literature do *not* serve as stringent tests of integrative hypotheses. Discussions of the potential of textual analysis methods are thus simply not the place to find appeals to stringency in the first place. But the repeated emphasis on the importance of experimental testing might well be read as an acknowledgment of the inherent necessity of assuring it, even if that work is pursued elsewhere.

Summing up, then—and, of course, unsurprisingly given that I hand-picked the case studies for precisely this purpose—it seems that literature-based discovery invokes all the constituent elements of integrative promise. But that wasn't our target here. It's equally clear that many other uses of textual analysis *won't* have any obvious connection to integrative promise, so no fully general argument for this conclusion can possibly succeed. My target is, instead, a broader one: what is it about textual analysis methods that makes their use particularly common in and pertinent for the life sciences?

As you might expect, my answer will nonetheless generalize from the relationship that we saw between integrative promise and textual analysis in these examples. I want to suggest here that the importance of integrative promise for the life sciences gives rise, more or less directly, to an significant part, at least, of the demand for textual analysis. Put differently, textual analysis is an especially suitable tool for helping us build explanations that exemplify integrative promise.

Recall the introduction of inter-domain integration in Chap. 2. I argued there that the life sciences are pressed by demands, not present to the same extent in other sciences, to integrate across disciplines. There is an increasing sentiment, as we've seen throughout the case studies in this book, that solutions to significant biological problems can only be found when we successfully combine knowledge arising at different scales, about different organisms, and developed in different disciplinary traditions. We can choose to respond to this state of affairs in a variety of ways. One such way, clearly, is the appeal to interoperable biological databases that I've already considered here; these databases are explicitly constructed so as to encourage the linkage of data at these different levels and the construction of integrative inferences. (Though, in the spirit of the critique I raised in Chap. 6, we should be careful to evaluate whether these are cases of mere data integration or, instead, the proposal of inter-domain-integrative explanations in my sense.)

But another such way appeals directly to the record of scientific publications. To put it simply, unstructured text is among the most important fundamental raw materials produced by the scientific process. Whether papers are written by biochemists, geneticists, or doctors, they end up (at least, in the early twenty-first century) as digitized records on centralized repositories, amenable to the tools of textual analysis. To be sure, and as any practitioner will gladly confirm (myself included), the analysis of natural language in unstructured text with the aim of arriving at novel insights is an extremely difficult undertaking. Our resources for doing this kind of work change rapidly. To take just one example, at the time of writing, the impact of large language models is only beginning to make itself felt in the scholarly process, and it is likely that technological advances will make the case studies that I have offered here seem quaint within only a few years.

The constant, I argue, is the role that scientific publications have, over the last several centuries, played in the dissemination of scientific knowledge.[8] As a kind of common denominator, any science that is deeply interested in questions of inter-domain integration will almost necessarily be drawn to tools that can create links between the texts produced by those different domains. Inter-domain integration thus seems to provide a strong incentive to analyze scientific texts. To an increasing degree, this fact is recognized by publishers, who have begun to make their content explicitly available for "text and data mining"—at least, to those researchers working at institutions who have purchased access rights to the content in the first place. Questions of a digital divide in research remain relevant, over a decade after the potential for such a corrupting influence on research was noted by scholars like danah boyd and Kate Crawford (2012).

So much for integration. The case for opportunism can obviously be made fairly quickly—setting aside the caveat about the expense of major journals, the knowledge already present in the scientific literature is effectively "free." At the very least, it doesn't require any further costly and perhaps epistemically risky experimental effort. This consideration of risk can be particularly powerful in contexts like the drug-development examples I have discussed here. As has been argued by, among others, both Alves et al. (2022) and Baker et al. (2018), previously approved drugs will readily succeed their phase-one clinical trials (which

[8] One might wonder, as some scientists already have (Bourne 2005), whether or not a future descendant of "scientific publication" will cease to have this unstructured textual component. I'm skeptical, but I think arguments in either direction engage in too much futurology to be interesting for my purposes here.

demonstrate that the compound is safe for human consumption in the first place), and will be more likely to successfully navigate the more important phase two and three trials (which demonstrate the efficacy of the medication to treat the condition at issue). Opportunism in the sense of reduction of these risks of failure will therefore be crucial in this context.

As already noted, there will be less to say in general about the relationship between textual analysis and either stringency or non-epistemic values. Of course, not every usage of textual analysis will be in the service of integrative promise. One might pursue some more local cases of integration where the stakes are lower, and one might pursue them without being particularly worried whether the knowledge that results provides a stringent test for that purported integration. (Perhaps you only are on the hunt for interesting proposals for ways in which you might spend the last few euros of your grant funding, even if the knowledge that results is likely to serve as little more than interesting trivia.) In such a case you'd lack both stringency and engagement with non-epistemic values; but in such a case you'd equally well not be pursuing integrative promise in the first place.

In short, then, I argue that there are good reasons to think that textual analysis will often support the hunt for integrative promise, especially because life scientists are particularly well positioned to develop inter-domain integration in an opportunistic context. As these compose half of the cluster of explanatory virtues that make up integrative promise, this gives rise rather naturally to a strong link between promise and textual analysis.

One more point should be raised before we conclude the descriptive portion of this discussion.[9] There is an apparent tension between the presentation of textual analysis methods in this chapter and the presentation of "theory-free" biology that we saw in Chap. 6. Couldn't we see textual analysis as providing more complex, nuanced, yet still data-driven knowledge of precisely the sort that would make it possible for us to avoid some of the objections that I raised earlier to the prospect of transcending theory?

I think this conclusion would be too hasty, but the parallel is interesting and instructive. First, to be sure, precisely the same problems of data integration and stringency arise in the case of textual analysis as arise in other big-data projects. Getting your hands on the right text, text that makes possible a real evaluation of a given hypothesis, is usually far from straightforward. This is yet another example of the importance, as Leonelli (2016) has noted, of practical know-how in the performance of these kinds of analyses. There is no "objective" measure of the fit of a dataset to a research problem; any attempt to justify one's choice in such a way can only be an attempt to hide a subjective choice on the part of the researcher.

Second, and relatedly, consider how these analyses are actually done. Once more, there is a tendency to see textual analyses as simply "reading off" their results from

---

[9] Thanks to an anonymous reviewer for raising this point, which I had embarrassingly failed to consider in an earlier version.

"the data," in such a way that no one could doubt, for instance, that a given kinase has (at the very least) an increased probability of interacting with p53 given that we detected a significant correlation between the two. After all, the existence of this correlation seems to be a "brute fact" about the scientific literature, regardless of one's theoretical presuppositions.

Leave aside, for the moment, the fact that any such correlation had to be confirmed biochemically before being accepted by the scientific community—casting doubt on the idea that this is really a "brute" scientific fact waiting to be found in the literature. (I'll return to this point in the next section.) Instead, consider now what kinds of decisions had to be made for us to even arrive at the observation of that correlation. As we saw just above, first a corpus had to be constructed: only some papers are going to be considered as "legitimate" sources in the context of this investigation, likely sorted by journal, but also potentially sorted by things like perceived quality.[10] Then, a number of further choices need to be made: texts must be cleaned, textual reference to the proteins at issue must be accurately detected (with outliers or noise separated and discarded), a network must be created (perhaps with certain thresholds for minimal levels of connection, or other, graph-theoretical cleaning methods applied, such as taking the largest connected component in order to avoid isolated "islands" in our network).

Each of these, then, is a paradigmatically subjective or free choice on the part of the investigator. Some might be set sociologically, by "best practices" in the field in which they work; some might be personal choices or the result of intuition. Some of these constraints, too, will be set by the theories in which we work: certain kinds of interactions, say, are ruled in or out ahead of time, because our extant theoretical knowledge tells us that they are particularly (un)interesting. In any case, the message is clear: we should resist the idea that even these tools are capable of straightforwardly "reading off" their results from the data. While the studies described in this chapter can be profitably read in dialogue with the nuanced reading of theory-free biology that I laid out in Chap. 6, they do not seem to be different in kind.

8.4 Using Textual Analysis for Promise

To close—and to anticipate the topic of the book's next and final chapter—I want to turn to a normative dimension. If textual analysis is *sometimes* useful for integrative promise, how could we construct our tools so as to enhance the possibility of

---

[10] This latter problem has been magnified by the spread in recent years of so-called "paper mill" journals, which will accept nearly any contribution from nearly anyone, provided that the article processing and publication fee is paid on time. It is already nearly impossible to say what "every journal article" on a subject would really mean, and it is equally true that even if we could make sense of such a claim, this isn't even the scope that we would want for our analyses.

arriving at that promise? This is no idle question for me, as I once maintained an active, publicly and freely accessible project based on a collection of around 1.9 million full-text journal articles (Ramsey and Pence 2016; Pence 2016). What could tools like these do, and do better, to support the pursuit of integrative promise, and how could other parts of the discipline help?

As already briefly noted above, questions of copyright and access loom large. The landscape improves every year, with more and more publishers deploying general-purpose licenses that look favorably upon this kind of work. Assembling a corpus, even an extremely large one, of biological texts is steadily becoming a simpler assignment.

But this is the easy part. The actual tools for analysis face a number of important, interlocking problems. Many of them require some knowledge of programming languages (the most popular at the moment for this purpose being Python and R). Already as early as 2010, biologists were recognizing that developing user-focused, comprehensible textual analysis applications would be one of the major challenges for the coming decade (Dai et al. 2010, p. 173). It's here that the real difficulty lies. First, simple technical requirements often mean that the "useful" and "usable" parts of digital applications become cumbersome, unwieldy, and outdated much faster than the data upon which they rely. After all, a given scientific article, once published, printed, digitized, and processed into a machine-usable form will—modulo changes in data representation formats and so on—be a reasonably static object. The interface that is needed to let users analyze that object, on the other hand, will likely radically and quickly change, as anyone who has used even "standard" software over the last twenty or thirty years can attest.[11] If "maintaining access" to a textual analysis resource includes not just ensuring that the data remains in a useful and usable format, but that the user interface remains up-to-date, accessible, and performs well, then such maintenance is a tall order. (It is also, as many digital humanists will lament, often an unfunded order.)

This kind of accessibility, however, is vital for realizing the integrative promise of textual analysis. After all, such analyses need to not only be usable, but to allow us to pose questions that might have been entirely unenvisaged at the time that the resource was created. To take just one example, the kinds of sequences of letters and numbers that make up gene and protein names in the molecular biology literature (p53 or NEK2 from our example, say) are precisely the kinds of tokens that might be discarded by an over-zealous attempt to automatically clean texts. (Indeed, they don't much look like words.) One could even have particularly good reasons for having removed them—for instance, if the corpus was initially assembled only for the purposes of analyzing prose, then any strings like "p53" present within it are likely errors in the data. But in making this choice, which might be less than fully transparent to the analyst and entirely invisible to the end-user, we have unknowingly built a wall that renders inter-domain integration impossible.

[11] Think of the difference—both in user experience and, equally if not more profoundly, in underlying code—between a text- or terminal-mode spreadsheet application like Visicalc, an early graphical spreadsheet like Lotus 1-2-3, and recent versions of Microsoft Excel or Google Sheets.

There is also much work to be done in improving the stringency of textual analysis itself—relying less, as in the case studies here, on subsequent experiment to "pick up the slack." How can we more clearly assess the quality or robustness of these analytical inferences? What *is* a stringent evaluation of the kinds of generalizations produced by textual analysis? What would best-practices for ensuring such stringent tests look like? We largely lack answers to these questions (Pence 2022b). By comparison to our colleagues in statistical analysis, for instance, textual analysis is less often grounded in solid mathematical proofs of the characteristics of our analytic methods.[12] To be sure, we'll never be able to arrive at "fully-formed" pieces of biological knowledge without empirical support. But much time and money could be better directed, I think, with the aid of a fuller understanding of stringency in this context.

More provocatively, our ability to respond to grand challenges with these tools is also often hampered by their monolingual nature (Leonelli has worried about the same consideration in the case of biological databases; Leonelli 2016, p. 56). Analyzing text in Arabic or Chinese is an entirely different proposition than restricting one's analysis to Western European languages, and this is already a significant difficulty compared to English-only corpora. Working with multilingual corpora poses serious technical and linguistic obstacles, especially if one hopes to analyze semantic content across such a corpus (Malaterre and Lareau 2022). But if, as is often the case, the problems that grand challenges confront us with have their first and most serious impacts on people in developing countries, then engaging with the science produced in those countries by those very scholars will be vital.

Integrative promise thus serves as an especially interesting lens through which to view the use of textual analysis in the life sciences. The fact that the life sciences do indeed seek integrative promise seems to give us an explanation for the more extensive use of these methods in these domains compared to the physical sciences. Textual analysis is particularly suitable for opportunistically securing vertical integration; finding it where integrative promise is often pursued is thus relatively unsurprising. We can also look outward, using integrative promise as a way in which to think about the future of textual analysis. As we have seen time and again, integrative promise is hard-won; developing these methods in specific directions so as to secure it is a worthwhile goal for future work.

References

Altman RB, Bergman CM, Blake J, Blaschke C, Cohen A, Gannon F, Grivell L, Hahn U, Hersh W, Hirschman L, Jensen LJ, Krallinger M, Mons B, O'Donoghue SI, Peitsch MC, Rebholz-Schuhmann D, Shatkay H, Valencia A (2008) Open access text mining for biology - the way forward: opinions from leading scientists. Genome Biol 9(Suppl 2):S7. https://doi.org/10.1186/gb-2008-9-S2-S7

[12] Among other technical reasons, this is in part due to the fact that very little in the statistics of natural language is normally distributed, limiting the use of tools with robust mathematical proofs of effectiveness.

Alves VM, Capuzzi SJ, Baker N, Muratov EN, Trospsha A, Hickey AJ (2020) Mining complex biomedical literature for actionable knowledge on rare diseases. In: Bizzarri M (ed) Approaching complex diseases, vol 2. Springer International Publishing, Cham, pp 77–94. https://doi.org/10.1007/978-3-030-32857-3_4

Alves VM, Korn D, Pervitsky V, Thieme A, Capuzzi SJ, Baker N, Chirkova R, Ekins S, Muratov EN, Hickey A, Tropsha A (2022) Knowledge-based approaches to drug discovery for rare diseases. Drug Discov Today 27(2):490–502. https://doi.org/10.1016/j.drudis.2021.10.014

Baker NC, Hemminger BM (2010) Mining connections between chemicals, proteins, and diseases extracted from Medline annotations. J Biomed Inform 43(4):510–519. https://doi.org/10.1016/j.jbi.2010.03.008

Baker NC, Ekins S, Williams AJ, Tropsha A (2018) A bibliometric review of drug repurposing. Drug Discov Today 23(3):661–672. https://doi.org/10.1016/j.drudis.2018.01.018

Bourne P (2005) Will a biological database be different from a biological journal? PLoS Comput Biol 1(3):179–181. https://doi.org/10.1371/journal.pcbi.0010034

Boyd D, Crawford K (2012) Critical questions for big data. Inf Commun Soc 15(5):662–679. https://doi.org/10.1080/1369118X.2012.678878

Boyle F, Sherman D (2006) Scopus® Core™: the product and its development. Ser Libr 49(3):147–153. https://doi.org/10.1300/J123v49n03_12

Chang H (2012) Pluralism in science: a call to action. In: Is Water H2O? Evidence, realism and pluralism. Springer, Dordrecht, pp 253–270

Choi BK, Dayaram T, Parikh N, Wilkins AD, Nagarajan M, Novikov IB, Bachman BJ, Jung SY, Haas PJ, Labrie JL, Pickering CR, Adikesavan AK, Regenbogen S, Kato L, Lelescu A, Buchovecky CM, Zhang H, Bao SH, Boyer S, Weber G, Scott KL, Chen Y, Spangler S, Donehower LA, Lichtarge O (2018) Literature-based automated discovery of tumor suppressor p53 phosphorylation and inhibition by NEK2. Proc Natl Acad Sci 115(42):10666–10671. https://doi.org/10.1073/pnas.1806643115

Cuatrecasas P (2006) Drug discovery in jeopardy. J Clin Investig 116(11):2837–2842. https://doi.org/10.1172/JCI29999

Dai HJ, Chang YC, Tzong-Han Tsai R, Hsu WL (2010) New challenges for biological text-mining in the next decade. J Comput Sci Technol 25(1):169–179. https://doi.org/10.1007/s11390-010-9313-5

Farrell MJ, Brierley L, Willoughby A, Yates A, Mideo N (2022) Past and future uses of text mining in ecology and evolution. Proc R Soc B: Biol Sci 289(1975):20212721. https://doi.org/10.1098/rspb.2021.2721

Henry S, McInnes BT (2017) Literature based discovery: models, methods, and trends. J Biomed Inform 74:20–32. https://doi.org/10.1016/j.jbi.2017.08.011

Hunter LE (2017) Knowledge-based biomedical data science. Data Sci 1(1–2):19–25. https://doi.org/10.3233/DS-170001

Klein U (2001) Paper tools in experimental cultures. Stud Hist Philos Sci 32(2):265–302

Kundu K, Darden L, Moult J (2021) MecCog: a knowledge representation framework for genetic disease mechanism. Bioinformatics 37(22):4180–4186. https://doi.org/10.1093/bioinformatics/btab432

Lean OM, Rivelli L, Pence CH (2021) Digital literature analysis for empirical philosophy of science. Br J Philos Sci. https://doi.org/10.1086/715049

Leonelli S (2016) Data-centric biology: a philosophical study. University of Chicago Press, Chicago

Malaterre C, Lareau F (2022) The early days of contemporary philosophy of science: novel insights from machine translation and topic-modeling of non-parallel multilingual corpora. Synthese 200(3):242. https://doi.org/10.1007/s11229-022-03722-x

Mongeon P, Paul-Hus A (2016) The journal coverage of Web of Science and Scopus: a comparative analysis. Scientometrics 106(1):213–228. https://doi.org/10.1007/s11192-015-1765-5

Pence CH (2016) RLetters: a web-based application for text analysis of journal articles. PLoS One 11(1):e0146004. https://doi.org/10.1371/journal.pone.0146004

Pence CH (2022a) How not to fight about theory: the debate between biometry and Mendelism in *Nature*, 1890–1915. In: Block AD, Ramsey G (eds) The dynamics of science: computational frontiers in history and philosophy of science. University of Pittsburgh Press, Pittsburgh, pp 147–163

Pence CH (2022b) Testing and discovery: responding to challenges to digital philosophy of science. Metaphilosophy 53(2–3):238–253. https://doi.org/10.1111/meta.12549

Pence CH, Ramsey G (2018) How to do digital philosophy of science. Philos Sci 85(5):930–941. https://doi.org/10.1086/699697

Przybyła P, Shardlow M, Aubin S, Bossy R, Eckart de Castilho R, Piperidis S, McNaught J, Ananiadou S (2016) Text mining resources for the life sciences. Database 2016. https://doi.org/10.1093/database/baw145

Ramsey G, Pence CH (2016) evoText: a new tool for analyzing the biological sciences. Stud Hist Philos Biol Biomed Sci 57:83–87. https://doi.org/10.1016/j.shpsc.2016.04.003

Rappaport N, Twik M, Plaschkes I, Nudel R, Iny Stein T, Levitt J, Gershoni M, Morrey CP, Safran M, Lancet D (2017) MalaCards: an amalgamated human disease compendium with diverse clinical and genetic annotation and structured search. Nucleic Acids Res 45(D1):D877–D887. https://doi.org/10.1093/nar/gkw1012

Searls DB (2001) Mining the bibliome. Pharmacogenomics J 1(2):88–89. https://doi.org/10.1038/sj.tpj.6500030

Soldatova LN, King RD (2006) An ontology of scientific experiments. J R Soc Interface 3(11):795–803. https://doi.org/10.1098/rsif.2006.0134

Soussi T (2014) The TP53 gene network in a postgenomic era. Human Mutat 35(6):641–642. https://doi.org/10.1002/humu.22562

Swanson DR (1986) Fish oil, Raynaud's syndrome, and undiscovered public knowledge. Perspect Biol Med 30(1):7–18. https://doi.org/10.1353/pbm.1986.0087

Swanson DR (1988) Migraine and magnesium: eleven neglected connections. Perspect Biol Med 31(4):526–557. https://doi.org/10.1353/pbm.1988.0009

Wishart DS, Knox C, Guo AC, Cheng D, Shrivastava S, Tzur D, Gautam B, Hassanali M (2008) DrugBank: a knowledgebase for drugs, drug actions and drug targets. Nucleic Acids Res 36(Suppl_1):D901–D906. https://doi.org/10.1093/nar/gkm958

Wishart DS, Feunang YD, Guo AC, Lo EJ, Marcu A, Grant JR, Sajed T, Johnson D, Li C, Sayeeda Z, Assempour N, Iynkkaran I, Liu Y, Maciejewski A, Gale N, Wilson A, Chin L, Cummings R, Le D, Pon A, Knox C, Wilson M (2018) DrugBank 5.0: a major update to the DrugBank database for 2018. Nucleic Acids Res 46(D1):D1074–D1082. https://doi.org/10.1093/nar/gkx1037

Chapter 9
Conclusion: Supporting the Search for Promise

Abstract In the book's conclusion, I turn toward the future, attempting to find ways that we can use careful consideration of the structure of integrative promise to offer recommendations for biological practice. If integrative promise is really as important to biology as I've argued here, then we should take it as a goal to cultivate each of its component parts. I begin by detailing a number of reasons for which interdisciplinary work is difficult to sustain in the contemporary academy, making inter-domain integration particularly challenging. I then explore the possibility of open science, particularly data reuse, for bolstering the prospects for opportunism. The kind of stringency that is required by promise is threatened, I claim, by the lack of publication of negative results; the same is true for the lack of engagement between science and ethics and policy for the resolution of socially important problems. I close with a brief summary.

Keywords Interdisciplinarity · Scientific training · Open science · Data reuse · Negative results · Science and values

A common refrain throughout this book, to which we returned at the end of the last chapter, has been the difficulty with which integrative promise is attained. As I have presented it here, integrative promise is a collection of explanatory virtues: explanations that embody it are genuinely horizontally and/or vertically integrative, they relate in the right ways to the enabling force of opportunism and the constraint of stringency, and they can help us to advance non-epistemic values, often via confronting grand challenges. It is, in some sense, impressive that contemporary biological science has the ability to produce such explanations in the first place and, equally, entirely unsurprising that this pursuit would be a tenuous and risky enterprise, the history of which is riddled with failed proposals and hype.

In this final chapter, I want to look toward the future. I'll take for granted that integrative promise is a desirable goal for biological explanations. Crises like climate change, biodiversity loss, zoonotic pandemics, and the prospects for the drug development pipeline are among the most important questions for contemporary science to tackle, and—at least if the arguments that I've detailed here, offered both

C. H. Pence, *Integrative Promise*, Synthese Library 499,
https://doi.org/10.1007/978-3-031-88471-9_9

by myself and by scientific practitioners, are correct—they will require explanations with integrative promise. What I want to consider now is a higher-level, normative question. What are some of the characteristics of contemporary biological science that contribute to making integrative promise so difficult to achieve, and how might we envision proposals for making it more likely to successfully produce promising explanations?

To unpack this question, I'll consider each of the four components of integrative promise in turn, raising for each a potential worry about how contemporary scientific practice might make us less likely to arrive at that explanatory virtue, and, finally, making some recommendations for how we might refine practice to raise our odds of accomplishing this goal.

9.1 Integration: Practicing Interdisciplinarity

The relationship between inter-domain integration and interdisciplinarity was already discussed at some length in Sect. 2.2. Additionally, a number of the examples that I introduced have made even more direct appeals to the importance of interdisciplinary study to their work. I extensively explored the integrative promise of systems biology in Chap. 6, and in exactly this vein, Jane Calvert and Joan Fujimura, drawing on numerous interviews with investigators in systems biology, have demonstrated the centrality of interdisciplinarity to this burgeoning field. As they put it, "one of the key features of the systems biology approach is its *inter*disciplinarity, which is said to be unavoidable, because the skills of many disciplines are necessary to solve the problems that are being raised by the field" (Calvert and Fujimura 2011, p. 156). Precisely one of the things that differentiates, in their view, systems biology from traditional molecular biology is its reliance upon these interdisciplinary methods. Sean Eddy (2005) has made the same point with regard to computational biology.

As I noted in Chap. 2, pinning down what this sense of interdisciplinarity is can be extremely difficult. We want a type of collaboration from which genuinely integrative explanations can emerge. Those explanations won't be the result of simply carrying theoretical or mechanistic claims from one field wholesale into the explanatory domain of another, in what Renate Mayntz has called *reductionist extension*:

> In [reductionist extension], a discipline and its explanatory claim[s] are extended into the phenomenal domain of another discipline. This takes place, for instance, when the chemist starts to analyze and explain biological phenomena or when the biologist tries to explain social features of modern societies as a consequence of biological, Darwinian evolution. (Mayntz 1992, p. 28)

If inter-domain integration, as I defined it here, is intended to capture a more robust relationship, in which explanations are produced that transform those that would be offered by any single domain, then this kind of reductionist extension simply won't do.

But it is precisely this interdisciplinary character—not in the superficial sense of the borrowing of some explanatory resources, but in the more comprehensive sense of transformative interdisciplinarity—that is also taken to pose a significant problem for at least some of the domains that we've discussed thus far. Returning to the case of systems biology, for instance, Calvert and Fujimura write, quoting a computer scientist whom they interviewed, that "you must not under-estimate the importance of culture in blocking interdisciplinary advances"; another biologist noted that launching an interdisciplinary center was "an enormous struggle" (Calvert and Fujimura 2011, p. 157).

At least four practical roadblocks seem to lie in the way of our ability to create this kind of interdisciplinarity. The first relates to individual training. While it is possible that the relevant kind of integration could be produced simply by forming teams composed of researchers trained in different fields, it seems to regularly be the case that this is not how real-world transformative interdisciplinarity has been developed. Eddy has argued, on the basis of experience in computational biology, that usually *individual* experience is required. "When I think," he writes, "of new fields in science that have been opened, I don't think of interdisciplinary *teams* combining existing skills to solve a defined problem—I think of single interdisciplinary *people* inventing new ways to look at the world" (Eddy 2005, p. 3).

There is thus a very concrete concern here, having to do with the skill-sets of particular practicing scientists. If Eddy is right, then we need to find ways to ensure that young researchers have enough familiarity with neighboring (or not-so-neighboring?) disciplines and their experimental and theoretical techniques to support interdisciplinary perspectives. Answering this question may have to do with the additional obstacles that stand in the way of such research.

Secondly, then, this sort of interdisciplinarity seems to require significant changes not just in our ways of pursuing scientific explanations, but moreover to the entire institutional organization of scientific training and departments in the contemporary university. Of course, this is not a novel insight. Perhaps the most vocal defender of this position has been Nicholas Maxwell, who has argued at length that the current structure of academic inquiry does not place us in a position to be able to respond to large-scale social problems that demand this kind of analysis (e.g., Maxwell 2014). But implementing any such changes will be a serious challenge in an environment where, in nearly every "traditional" discipline, program coordinators constantly search for more course hours for technical subjects.

Some promising models—for instance, the creation of problem-based and project-based learning in the sciences, or the cultivation of research centers that sit between groups of thematically related scientific and humanities departments (e.g., surrounding questions in neuroscience research or climate change)—have already been developed, and it will be important to see how they unfold in the future. But the rather stubborn reality will remain that, in most cases, students still need a diploma carrying a single departmental affiliation that is intelligible to future employers, and universities are largely built around the model of hierarchically organized departments.

We should then consider, as an alternative, how this work might look in the absence of these kinds of major structural reforms—more specifically, via constructing interdisciplinary teams that bring together people trained in "traditional" approaches to their respective disciplines. As already noted, however, if we want to construct new, collaborative models of interdisciplinarity that do rely upon teams, we must come to terms with the frequent failure of this model in the past. Why has the team-based approach not seemed, *de facto*, to be a driver of the generation of promising explanations? In part, and unfortunately, this is because many of the same challenges that we just identified for a thoroughgoing, individual interdisciplinarity will hold even in this more limited case. As philosophers of science who would like to work with practicing scientists know well, recruiting scholars—even those who might be excited by the integrative promise of a potential line of research in the abstract—who are comfortable working across disciplinary boundaries can be a difficult proposition. This is so, I think, as a result of at least two further problems with transformatively interdisciplinary work.

Harkening back to the last chapter, the third roadblock has to do with publication and credit. Scientific journals are still, in the vast majority of cases, disciplinary objects.[1] Publishing in the "wrong field" will, at best, be seen as an interesting side-project by hiring and promotion committees. How, then, can we ensure that scholars doing work with integrative promise will be adequately credited for their research? Does every integrative pursuit require the creation of novel disciplinary, professional, or publication venues—as has happened, for instance, in the cases of systems biology and computational biology?

This point runs headlong into concerns about open science, which will be the topic of the next section. As the publication landscape changes more broadly, moving from a small number of high-prestige journals to a radically open environment of preprints shared via social media and a much faster rate of knowledge circulation, these questions about credit will only become more poignant. How can we measure the "impact" of work distributed in these ways? Should we? Do those measures adequately take into account the breadth of the impacts that promising work might have, when they could be at the social or cultural level, and could lie outside the scope of traditional academic centers of power? Schemes of researcher evaluation and scientometrics will likely need to very seriously adapt in response to this challenge; such adaptation would, hopefully, itself help to support the pursuit of integrative promise.

The fourth and final point turns to the medium-term career consequences of performing this sort of integrative work. If publishing in unusual venues, working in unusual institutional structures, or obtaining unusual training is seen as a hindrance to the standard routes of career development, then asking a doctoral or post-doctoral researcher to contribute to such projects is a gamble—if the integrative promise

[1] For a particularly illuminating broad history of scientific journals, see Csiszar (2018).

pays off, then they may be seen as a successful and innovative scholar with some prestigious articles to their name. But if it fails to do so, then they may instead be left with training, experience, and outputs that their profession doesn't know how to evaluate, either because they are too integrative or because they involve overt non-epistemic value commitments that some might see as suspect.

As the academic job market becomes progressively more ruthless, faculty members need to bear in mind our moral responsibility toward our trainees: we cannot ask people to participate in research that could, in essence, be a career-killer. If we want to encourage a generation of young researchers to pursue integrative promise, then, we must find ways to ensure that their academic career path will be no worse off (indeed, quite the contrary) as a result. One remedy relates to long-standing calls for "public outreach" activities—a category into which at least some transformative interdisciplinary work is often lumped—to become a more central part of the careers of scientists (see, e.g., Varner 2014; Graur 2018). But this only solves part of the problem. Much of what is produced in these contexts will consist of perfectly normal scientific work that happens to fall outside of one's "home community." Integrating these projects and publications into scholarly trajectories is a more challenging endeavor.

In short, it seems as though the current structure of education, training, careers, advancement, and publication—surely a complexly interwoven set of problems—does not encourage scholars to develop the kinds of interdisciplinary skills that foster integrative promise. It is noteworthy that some of our best examples of the pursuit of promise come from "non-traditional" institutions: the European Molecular Biology Laboratory offers some degree of "insulation" of its members from the concerns of academic departments, as does the fact that the practice of natural history is often performed in contexts like museums and field stations where the metrics for career advancement are constructed and understood at least somewhat differently.

As I think this short inventory of obstacles has demonstrated, there will be no simple fix here. If inter-domain integration requires, as I've argued, genuine transformative interdisciplinarity, then it is at odds with a number of deep-seated features of the contemporary academic enterprise. Without concerted efforts to lower these barriers, the development of such integration may remain somewhat fortuitous.

Finally, it is significant that the four roadblocks (and potential solutions) that I've explored here largely remain in the domain of the practical. One might well worry that integration would be prevented by more serious, fundamental, and perhaps unresolvably incommensurable differences between disciplines. Indeed, disciplinary training instills a whole host of both explicit and tacit practices, norms, and so on. These perspectives might simply block any possibility of performing genuinely interdisciplinary work before it even gets off the ground. There's some truth behind this concern, and despite my optimism in this section, it's important to recognize the limits—and the difficulty—of these practical interventions. Nonetheless, I think that the importance of this kind of work and the potential for reform means that we are obligated, as philosophers of science, to explore these epistemic and practical possibilities.

9.2 Opportunism: Open Science

Let's turn now toward ways in which we might facilitate opportunism. To do so, I want to consider discussions of data reuse developed in the context of the contemporary open science movement. This is something of an unusual approach to take with respect to the philosophical and scientific literature on open science. Pasquetto et al. describe reuse as a more recessive feature of the literature as a whole, writing that "data sharing practices, especially motivations and incentives, have received far more study than has data reuse, perhaps because of the array of contested concepts on which reuse rests and the disparate contexts in which it occurs" (Pasquetto et al. 2017, p. 1).[2]

For Pasquetto et al., this raises a serious issue: despite longstanding claims that the reuse of data—and the opportunistic research programs that it can enable—is among the central motivations for open science (e.g., Molloy 2011), they note that many research questions about data reuse are not yet well understood. Even if data are, strictly speaking, available, this is only the tip of the iceberg. As they go on to argue,

> Reusing a single dataset in its original form is difficult, even if adequate documentation and tools are available, since much must be understood about why the data were collected and why various decisions about data collection, cleaning, and analysis were made. (Pasquetto et al. 2017, p. 4)

Unresolved research questions only accumulate from there. Perhaps most importantly, the premise that data reuse is indeed a practice that will permit opportunism often remains unexplored. But it is assuredly the case that the choice between the reuse of an existing dataset and the preparation of a new one will involve a variety of trade-offs—trade-offs that have not yet been described in detail.

Every dataset comes with underlying assumptions about the object of study. Certain pieces of metadata will be chosen for retention, others will be ignored; variables will be represented as continuous, discrete, or categorial; some might use preexisting interoperable standards, others could follow locally created and documented standards, and yet others might conform to no standards at all, products of dynamic and flexible classifications built in response to changing needs. Every project has to make often excruciating decisions concerning its scope, as researcher time, server space, and grant funding are not unlimited resources. No reused data will thus be "made to measure," and the process of adapting a dataset for

[2] An important exception to this trend is Sabina Leonelli's recent work on the philosophy of open science, which arrived too late for me to be able to extensively consider it in writing this section (Leonelli 2023).

reuse will require inherent compromises. To some extent, the dynamics of that adaptation process will be very local and will generalize only with difficulty. But more detailed descriptions of their basic features would enable researchers to better assess the opportunism of a potential line of further research, or the extent to which opportunism has been privileged against other explanatory virtues (especially stringency) in a given explanation.

Equally important, and equally poorly understood, is the impact of preparation for reuse on the data collection process in the first place. Decisions to anticipate reuse will alter the future trajectory of scientific observation. Are these changes for the better or for the worse? Some evidence seems to indicate that data collected with reuse in mind are not just more functional for other teams that might engage in that reuse, but also for the very groups that produced the data in the first place, likely because higher quality metadata means that data producers can "reuse" their own work more effectively (Goodman et al. 2014). But again, given the complex trade-off structure here and the multiple demands on the time of a scientific group, this connection is not a necessary one.

As is often the case in digital projects of any stripe, the perfect can certainly be the enemy of the good.[3] If too much effort is put into finding common standards or ontologies, preparing data categories, and so forth in advance, the project might never get off the ground, or its costs might be massively amplified. But if not enough such effort is put in ahead of time, the technical debt—that is, the set of expensive technical choices made early in the lifespan of a project that are now "baked into" the project's very infrastructure—will accumulate, and changing those fundamental decisions long after the project's debut may become technically or financially impossible. To that end, Goodman et al. (2014) recommend targeting *a particular level* of data reuse ahead of time: envisioning potential audiences and their needs, with the idea in mind of later using this as a yardstick for directing reuse-enabling work.

A final important worry about data reuse in the context of open science concerns the relationship between data reuse and open science's political and ethical commitments. As has been indicated by Leonelli (2023), open science has a deeply political orientation, as it lauds a particular kind of openness, taken to be morally and socially beneficial and undergirded by free access to knowledge and the sharing of (often publicly funded) data. But yet again, these kinds of political commitments often go unquestioned.[4] Does data reuse actually encourage ethically superior science? In what way does it do so?

[3] The version of this adage most common in the computer programming community is that "premature optimization is the root of all evil" (Knuth 1974, p. 268).

[4] Leonelli argues at length that an emphasis on sharing and standardization could potentially allow problematic kinds of extractive practices to victimize the Global South, denying communities' agency to control the distribution and fate of locally generated scientific knowledge or objects of study.

With the resources of integrative promise, we can recast this question and make it more precise. Opportunism is only one of the multiple aspects that constitutes promise. I've argued that it's frequently appealed to in the context of promise precisely because, among other reasons, scientists are looking to raise the odds that a particular integrative enterprise will be fruitful (as relying on previously successful scientific investigation is one way in which to do so), and they are often hoping for answers at least reasonably quickly, given the social and cultural urgency of the grand challenges that promising explanations are often supposed to aid us in resolving.

These political overtones surrounding "openness" in open science push us to think about further reasons for supporting opportunism. It seems true that in an environment of infinite scientific time and funding, opportunism would not be a part of integrative promise; we could simply pursue any and all research possibly relevant to our question. But the real-world limitations of the scientific process come with real-world responsibilities to optimally invest our limited resources. Data reuse thus itself takes on ethical overtones: faced with grand challenges and the difficulty of developing integrative approaches, one might argue that we don't have the luxury *not* to reuse data, when this reuse can be appropriately balanced against its potential downsides.

While data reuse (and hence opportunism) is therefore a crucial component of the motivation for open science, the theoretical and empirical links between open science and opportunism are often underexplored, which makes it difficult for us to evaluate our normative obligations to pursue opportunistic projects. More work shedding light on these connections would assuredly be welcome, particularly in order to give scientists actionable best practices that can optimize the potential of opportunism (e.g., Goodman et al. 2014; Munafò et al. 2017).

9.3 Stringency: Scientific Publication and Negative Results

The idea that scientific publication could be biased by the literature's incentives to publish statistically significant results (however minimally significant) and ignore negative results (however decisively negative) is one of the oldest tenets in what might now be termed scientometrics. First described by Robert Rosenthal (1979), the "file drawer effect"—so-called after the myriad negative results feared to be filling researchers' filing cabinets—has since been refined in a number of ways, perhaps most celebrated among them the more complete Bayesian model of the probability that a given published result is true developed by John Ioannidis (2005).[5]

[5] Even earlier, Peter Medawar worried that the scientific paper as a publication format "misrepresents the processes of thought that accompanied or gave rise to the work that is described in the paper" (Medawar 1963, p. 42), a related problem that brings me toward issues of the purpose of scientific publication itself that colleagues and I have considered elsewhere (Lean et al. 2021).

The basic idea is simple, at least in the extreme.[6] Imagine that numerous labs full of scientists are testing an explanation that is in fact incorrect. By the simple luck (or lack thereof) of false positives, some labs (given a completely uniform distribution, one in twenty) will produce a positive result with $p < .05$. Of course, their competitors will have produced mountains of significantly more convincing negative data. But the positive result, especially if it supports an unconventional or surprising hypothesis, is more likely to land a splashy publication in an important journal, while the negative results are doomed to obscurity. In an echo of Swanson's epistemology of the medical literature in Chap. 8, the scientific community "knows" that this result is false—a synthetic evaluation of the data would clearly indicate it—but that knowledge is epistemically inaccessible to each of the community's members taken independently.

Why is the file-drawer effect important for integrative promise? Let me start by oversimplifying the point and then add back some missing detail. My aim in highlighting stringency as one of the constituents of integrative promise is, in part, to emphasize the difficulty in arriving at well confirmed integrative explanations of complex biological phenomena. It will often be the case that an integrative explanation under investigation does not, in fact, successfully explain the phenomena at issue. Finding the right kinds of tests—tests that are sufficiently stringent—for those explanations is no small problem, and so it's not implausible that methods will often be shared across competing labs. Taken together, it seems as though we will frequently be in precisely the position that the file drawer effect warns us against: that is, multiple research groups pursuing an ultimately negative result (i.e., arriving at the conclusion that a given integrative explanation fails), and largely using similar methodologies and asking similar questions. We thus place ourselves at a very significant risk that strong refutations of candidate integrative explanations will wind up buried in the file drawer. Integrative promise seems particularly prone to stringency failures of this sort.

This intuitive conclusion is borne out by a number of formal results. When we perform studies with lower prior probabilities, the model of Ioannidis (2005, pp. 698–699) shows us that the resulting probability that a published finding would be true becomes radically more sensitive to the model's other parameters: the statistical power of our experiments, the number of studies that are performed, and what he calls "bias," or the probability that an inaccurate result will be reported as an accurate one.[7] (Notable as well, for reasons that should be clear from our extreme example above, the relationship with the number of studies performed is in fact inverted—because of the file-drawer effect, the *more* laboratories are working

[6] Of course, real-world cases will be less extreme than this. The model of Ioannidis (2005) adds sophistication by introducing explicit parameters for the prior probability that the result is true, as well as the Type-I and Type-II error rates.

[7] The relationship between this resulting probability and the number of replications performed of a given study is also important (Moonesinghe et al. 2007), but getting into the discussion concerning replication would take me too far afield here (though I discuss preregistration at some length in Pence 2022).

on the problem, the *more* likely it is that a published result will be false!) One might then worry that, in the context of integrative promise, the distribution of these incorrect results would be magnified through processes like the Zollman effect, by which false results can become entrenched in highly interconnected epistemic communities (Zollman 2012).

What kinds of changes might enable us to have more confidence in stringent tests in an environment where the file-drawer effect has such potentially harmful effects? First, we can join a growing chorus of commentators arguing for the rethinking of scientific publication. New models for journals explicitly dedicated to the publication of negative results, such as the *Journal of Negative Results* in ecology and evolutionary biology (Kotze et al. 2004) or the more general *Journal of Trial and Error* (Devine et al. 2020), could help to encourage changes in the broader landscape of scientific publication. As with other institutional and cultural changes that I've considered above, though, this will be a slow and difficult undertaking, and it will have to navigate questions of credit and career advancement, as well as an intrinsic bias in scientific practice in favor of novel, innovative discovery and against replication of extant studies or the publication of null results.

As a recent European Union report has noted, funding agencies also have a role to play in these discussions, as they can "make clear that negative results and the verification of earlier findings are valuable" (Guédon et al. 2019, p. 47). If the standard procedure for developing a large-scale research project included the proposition of a plan for replicating positive results in nearby areas (very plausibly motivated as a way to increase the support for the results that the project would generate) and the provision of explicit arrangements for publication and distribution of negative results (perhaps including as well a commitment to the careful exploration of relevant negative results already published), we might see significant gains in stringency in a relatively short time-frame.

But I also think that we can see in this another benefit of looking at the full pattern of integrative promise for understanding the relationship between stringency and the other explanatory virtues invoked here. The importance of the publication and accommodation of negative results will doubtless vary, and, as already noted, explanations with integrative promise seem to be especially vulnerable to the file-drawer effect. But such explanations also have other virtues, virtues that might drive faster cultural change. Institutional resources are often mobilized around grand challenges that are absent in other parts of the scientific enterprise, both because of the social and ethical importance of those problems and because of the scope of integrative scientific study. Those organizations—for instance, the IPCC, the IPBES, the Human Genome Project, and so on—will thus have a sizable influence in encouraging more stringent theory evaluation, to disincentivize hype, and to push for more adequate management of negative results and study power. The tight connection to social and ethical values in this kind of work might itself play a role in motivating changes in behavior in the scientific community.

To be sure, the defense of integrative promise is not the only, or perhaps even the most persuasive, reason for supporting these kinds of changes to scientific practice. But recognizing that these explanations are both particularly difficult to derive and particularly vulnerable to worries like the file-drawer effect should prompt us to approach integrative promise with unusually high standards of care and methodological rigor. In a word, they argue yet again for the importance of stringency.

9.4 Science and Non-epistemic Values

Finally, let's consider the ability of the scientific community to attack socially and ethically relevant questions. Some of this has already been covered under discussions of interdisciplinarity earlier: to be sure, an important aspect of what makes grand challenges unusually daunting is their reliance on insights from other fields—potentially including a vast array of knowledge from multiple scientific communities as well as the social sciences and humanities.[8]

Equally clear is that scientists are very rarely given the chance to be involved with societal or ethical consideration of their work. I regularly teach students in scientific fields who tell me that the few weeks of philosophical course material that they receive are often the only time in their entire training that they have to step back, slow down, and weigh how to evaluate science from this sort of broader perspective. In that sense, it seems that just as integrative promise drives science to be more aware of these interactions between scientific pursuits and social and ethical values, scientists themselves are given increasingly fewer chances to develop the tools that they need to do this kind of work.

The question of the proper role for ethical values in science, or even the proper way in which scientists should be involved in social or ethical value debates, is now the subject of a literature so vast that it is impossible for me to do justice to it in the span of one section of a short conclusion. At the risk of offending several dozen colleagues who work in the area by omission, let me take a portion of Heather Douglas's recently published Descartes Lectures as a lens. Picking up midstream, she argues that today's professional scientists "are only experts in some areas of investigation, and the knowledge explosion that characterizes the past century demands specific expertise for specific problems" (Douglas 2021, p. 69). This expertise, however,

> poses a tremendous challenge to democratic governance. We need experts to inform our governance decisions, but we also need those decisions to be democratically accountable. [. . .] What does this mean and what *should* it mean in practice? We need governance to be assessable by the public, because that is who will make the recall decisions for the government. And we need those in power to be able to assess and use expertise in a

[8] This relationship is well known in the literature; Jan Schmidt (2008) has noted that part of interdisciplinary problem framing or problem perception results from a shift toward problems that are inherently defined outside the structure of single disciplines, often precisely because they are socially or ethically relevant.

> transparent way, so the public can assess what those in power are doing, particularly in response to the available expertise. (Douglas 2021, pp. 69–70)

I think it is no exaggeration to say that this cluster of questions is one of the most important of our age. As we saw in the response to the COVID-19 pandemic, just what it means for political decisions to "follow the science" is deeply unclear. Even if, as was largely the case, there is near-complete scientific consensus about a number of important empirical facts and relevant responses, two important constraints ensure that decision-making will remain complex. First, we often operate in environments of genuine scientific disagreement (what are the best epidemiological models for understanding the spread of SARS-CoV-2?), of lack of knowledge (what will be the economic and mental-health impacts of a massive shift to remote work?), or of limited resources (as in the case of the unavailability of high-quality masks in many places at the outbreak of the pandemic).

Moreover, and perhaps more significantly from a grand-challenges perspective, "following the science" can be, and has been, used as a way for democratic decision-makers to shirk their responsibilities. After all, "the science" will not tell us what we should value or, more practically, what levels of risk are socially or morally acceptable. Whether in the case of our response to COVID-19 or our (in)action in response to climate change, to deploy comprehensive political initiatives under the guise that they are straightforward logical consequences of scientific inquiry is a dangerous and inherently fallacious enterprise.

We thus need a multifaceted and carefully articulated model of the interaction between democratic governance and science, one that recognizes (1) the importance of the kind of knowledge generated by promising explanations to the management of the grand challenges that they are intended to solve; (2) the central function fulfilled by democratic and accountable decision-making in the resolution of social problems; and (3) the ineliminable role that ethical and social value judgments play in scientific practice (Longino 1990; Kitcher 2001; Douglas 2009; Biddle 2013; Brown 2013; Elliott and Richards 2017; Douglas 2021).

What does the public have the right to demand from its scientific experts? Douglas points to several answers to this question. First, they "should be able to give an account, an explanation, for why they think what they think" (Douglas 2021, p. 74). After all, anyone can pretend to give a scientific explanation—ten minutes on social media will convince most people of that—but only genuine competence offers adequate grounds for informed judgment. "But," she writes, "also on display should be the uncertainties and crucial points of judgment (particularly where other experts might judge differently)" (Douglas 2021, p. 74), as well as "the social and ethical values that shape expert judgment," including "the risks they are concerned about if they make an inaccurate claim" (Douglas 2021, p. 75).

Then we turn to the question of accountability for these decisions. Douglas parses this into several separate axes. First is the accountability to other members of the scientific community for the "reasonableness and accuracy of an expert's claims" (Douglas 2021, p. 76). This accountability is vital precisely because it is only those fellow scientists that are in a position to evaluate claims within their domain; it is unreasonable (if not dangerous) to expect members of the general public to be able to judge professional scientific conclusions. But the ability of the scientific community to hold itself accountable in such ways is not guaranteed. Whether in the case of the failure of accurate risk assessment prior to the L'Aquila earthquake in Italy (Douglas 2021, pp. 76–78) or the publication of modified avian flu strains with potential biosecurity implications (Evans et al. 2022), the default reaction of the scientific community is often to "defend science" from "meddling" by outsiders.

Second, experts need to be responsible for making clear to the general public the value judgments that inform their research. This will be hard work: not only is making one's own value commitments explicit a difficult endeavor, but—again, because of the high levels of training required to understand scientific studies—other members of the scientific community will likely need to verify that research indeed exemplifies (or fails to exemplify) the values that researchers claim for it.

But all of these avenues for the evaluation of science in a democracy are essential. As Douglas puts it, such a system

> should keep us from sliding into technocracy. We do not have to simply "trust the expert." We can demand to know how why they think what they think, see what their fellow experts think, and assess whether the values expressed are those we agree with. (Douglas 2021, p. 93)

To be sure, this is only part of the story. We still need to think about the obligations of the general public—for Douglas, this primarily takes the form of improved, bidirectional science communication that is able to better involve it in the scientific community, though exploring this facet of her position would lie far beyond my scope. We also still have practical questions of implementation to resolve. How could we actually put these kinds of systems into practice? What would the role of governments, funding agencies, and other civil society organizations like national academies be in that deployment?

I don't have answers to those questions. But I think that the introduction of integrative promise helps us bring them more clearly into focus. Put differently, one of promise's main advantages is that it lets us see how the evaluation of particular biological explanations can involve a variety of aspects, from narrowly drawn epistemic concerns, to broader explanatory virtues like stringency, to pragmatic factors like opportunism, to non-epistemic value commitments like the resolution of grand challenges, and even—in order to be sure that research with such non-epistemic value commitments is adequately responsive to society—questions about the governance of the scientific process.

9.5 Looking Forward

Let's end where we began. As I noted in the preface, my goal in writing this book was not to develop a novel theory of explanatory virtues in the life sciences, nor to convince you that extant such theories are radically flawed. I have instead worked to develop a tool for philosophical analysis—to pick out a pattern of invocation of these explanatory virtues, give that pattern a name, and show that investigating biology through this lens is a fruitful way to understand scientific practice.

Of course, it was never my intent to argue that *every* biological explanation exemplifies integrative promise. Much scientific work is, and is intended to be, incremental—"normal science" in Kuhn's sense. What matters for my purposes here is that this pattern does seem to appear often enough to illuminate a number of cases across the life sciences. In short, our philosophical toolkit is better off containing it than without it.

Battle-hardened philosophers of science may be wondering where to find the book's polemic. Who are my argumentative targets? I confess that, to my knowledge, there are none. The most devastating potential objection to what I've presented here would simply be that the reader finds no interest in the pattern that I've pointed out—that they believe that these explanatory virtues are best considered separately, and that the instances I've demonstrated of their coming together are no more than coincidence. I certainly disagree. I've tried both to show that these virtues are interconnected in important and non-trivial ways, and to show how an understanding of a scientific development in terms of integrative promise is more enlightening, more provocative, or richer than one that takes its components separately. But insofar as I've hoped to present a novel and useful implement for our toolbox of analysis in the philosophy of biology, the worst fate for a tool is to languish in perpetual inutility. I can only hope that others will find integrative promise as illuminating a framework for understanding the life sciences as I have.

The proof of that, though, will have to wait for the future. (I will resist the temptation to consider the integrative promise of integrative promise.) I want to close by considering just one more advantage of adopting this framework, on a somewhat personal and undeniably self-indulgent note. In the last several years, whether in pedagogical contexts, in professional contexts like the International Society for the History, Philosophy, and Social Studies of Biology (ISHPSSB), or in interdisciplinary contexts like the EMBL or the Académie Royale de Belgique, I've been struck by the crucial role that the life sciences are taken to play, time and again, in some of the most vital problems that face human society in the twenty-first century.

I think that philosophers of biology owe it to ourselves to avoid falling into either of two fairly attractive extremes. First, we might just dismiss all such invocations of integrative solutions to socially relevant problems as unmitigated hype, the kind of language that one puts in one's grant proposals to sell a workaday research project on *in vitro* molecular biology that will produce incremental advances, at best. But second, one might be carried away by that same hype, looking, as Michel Morange has poignantly cautioned us against, for "a biological solution to a problem whose origin is not necessarily biological, or even a biological solution to a problem which, even though it is biological, could be solved in non-biological ways. For the future of humanity is not biological" (Morange 2001, p. 33, my translation).

I turned to integrative promise precisely because I wanted to find both a term and a philosophical framework for analyzing cases like those presented here that avoided these two extremes. Looking at the kinds of explanatory virtues to which biologists are actually appealing in real-world cases can let us deploy the tools that we've refined over decades of work in philosophy of science, while at the same time applying those tools to explanations that might move us toward solutions to significant problems. If we can bring philosophical clarity and rigor to the very locations in which biology engages with socially and ethically important challenges, I think we can reinforce our role as scholars who explore the "complex and often troubled relations between science and society" and "make this relationship work" (Stotz and Griffiths 2008, p. 44).

Moreover, as we saw in several of the examples in the first few chapters, I don't think this pattern of explanatory virtues is entirely novel to the second half of the twentieth century. The application of the philosophy of science tools that I've used here, however, has been somewhat disfavored in the study of the history of science, largely out of well-founded worries that an overly restrictive pursuit of unified explanatory models does violence to the historical record (Galison 1988). While I share these worries, I don't want to throw the baby out with the bathwater. Looking at the ways in which Darwin tries to balance opportunism and stringency, I think, is every bit as illuminating as investigating the detailed development of his young theory of natural selection in his notebooks over the course of the 1830s (Ospovat 1981; Hodge 2009). Integrative promise can give us a concrete and clear path to explore the historical impact of philosophical positions, in a manner that, I hope, effectively generalizes while leaving enough room to let the details of particular historical cases matter.

And lastly, I also hope that raising these points from a philosophical perspective can be useful to our colleagues in the biological sciences. As we've seen in this final chapter, while the pursuit of integrative promise is certainly difficult, it doesn't rely only on random insight; various kinds of structural reforms could certainly allow us to better pursue integrative promise. I hope that practicing scientists will recognize something of their own motivations or evaluations in this presentation, and that being able to identify and argue for integrative promise explicitly could help provide them leverage to effect just the kinds of change that I've argued for here.

References

Biddle J (2013) State of the field: transient underdetermination and values in science. Stud Hist Philos Sci 44(1):124–133. https://doi.org/10.1016/j.shpsa.2012.09.003

Brown MJ (2013) Values in science beyond underdetermination and inductive risk. Philos Sci 80(5):829–839. https://doi.org/10.1086/673720

Calvert J, Fujimura JH (2011) Calculating life? Duelling discourses in interdisciplinary systems biology. Stud Hist Philos Biol Biomed Sci 42(2):155–163. https://doi.org/10.1016/j.shpsc.2010.11.022

Csiszar A (2018) The scientific journal: authorship and the politics of knowledge in the nineteenth century. University of Chicago Press, Chicago

Devine S, Bautista Perpinyà M, Delrue V, Gaillard S, Jorna TFK, Meer Mvd, Millett L, Pozzebon C, Visser J (2020) Science fails. Let's publish. J Trial Error 1(1). https://doi.org/10.36850/ed1
Douglas HE (2009) Science, policy, and the value-free ideal. University of Pittsburgh Press, Pittsburgh
Douglas HE (2021) The rightful place of science: science, values and democracy. The 2016 Descartes Lectures. Consortium for Science, Policy & Outcomes, Tempe
Eddy SR (2005) "Antedisciplinary" science. PLoS Comput Biol 1(1):e6. https://doi.org/10.1371/journal.pcbi.0010006
Elliott KC, Richards T (eds) (2017) Exploring inductive risk: case studies of values in science. Oxford University Press, Oxford
Evans NG, Selgelid MJ, Simpson RM (2022) Reconciling regulation with scientific autonomy in dual-use research. J Med Philos 47(1):72–94. https://doi.org/10.1093/jmp/jhab041
Galison P (1988) History, philosophy, and the central metaphor. Sci Context 2(1):197–212
Goodman A, Pepe A, Blocker AW, Borgman CL, Cranmer K, Crosas M, Stefano RD, Gil Y, Groth P, Hedstrom M, Hogg DW, Kashyap V, Mahabal A, Siemiginowska A, Slavkovic A (2014) Ten simple rules for the care and feeding of scientific data. PLoS Comput Biol 10(4):e1003542. https://doi.org/10.1371/journal.pcbi.1003542
Graur O (2018) Education and public outreach as an integral part of a scientist's career. Am J Phys 86(10):725–726. https://doi.org/10.1119/1.5052428
Guédon JC, Jubb M, Kramer B, Laakso M, Schmidt B, Šimukovič E (2019) Future of scholarly publishing and scholarly communication. Report of the Expert Group to the European Commission, European Commission, Brussels
Hodge MJS (2009) The notebook programmes and projects of Darwin's London years. In: Hodge MJS, Radick G (eds) The Cambridge companion to Darwin, 2nd edn. Cambridge University Press, Cambridge, pp 44–72
Ioannidis JPA (2005) Why most published research findings are false. PLoS Med 2(8):e124. https://doi.org/10.1371/journal.pmed.0020124
Kitcher P (2001) Science, truth, and democracy. Oxford University Press, Oxford
Knuth DE (1974) Structured programming with go to statements. Comput Surv 6(4):261–301
Kotze DJ, Johnson CA, O'Hara RB, Vepsäläinen K, Fowler MS (2004) Editorial: the Journal of Negative Results in Ecology and Evolutionary Biology. J Negative Results 1. https://www.jnr-eeb.org/index.php/jnr/article/view/1002
Lean OM, Rivelli L, Pence CH (2021) Digital literature analysis for empirical philosophy of science. Br J Philos Sci. https://doi.org/10.1086/715049
Leonelli S (2023) The philosophy of open science. Cambridge University Press, Cambridge. https://doi.org/10.1017/9781009416368
Longino HE (1990) Values and objectivity. In: Science as social knowledge: values and objectivity in scientific inquiry. Princeton University Press, Princeton, pp 62–82
Maxwell N (2014) How universities can help create a wiser world: the urgent need for an academic revolution. Societas, Exeter
Mayntz R (1992) The influence of natural science theories on contemporary social science. In: Dierkes M, Biervert B (eds) European social science in transition: assessment and outlook. Campus Verlag, Frankfurt, pp 27–79
Medawar PB (1963) Is the scientific paper a fraud? Listener 70:377–378
Molloy JC (2011) The open knowledge foundation: open data means better science. PLoS Biol 9(12):e1001195. https://doi.org/10.1371/journal.pbio.1001195
Moonesinghe R, Khoury MJ, Janssens ACJW (2007) Most published research findings are false–but a little replication goes a long way. PLoS Med 4(2):e28. https://doi.org/10.1371/journal.pmed.0040028
Morange M (2001) L'eugénisme aujourd'hui. In: Iacub M, Jouannet P (eds) Juger la vie: les choix médicaux en matière de procréation. Éditions La Découverte, Paris, pp 16–34
Munafò MR, Nosek BA, Bishop DVM, Button KS, Chambers CD, Sert NPd, Simonsohn U, Wagenmakers EJ, Ware JJ, Ioannidis JPA (2017) A manifesto for reproducible science. Nat Hum Behav 1:0021. https://doi.org/10.1038/s41562-016-0021

Ospovat D (1981) The development of Darwin's theory: natural history, natural theology, and natural selection, 1838–1859. Cambridge University Press, Cambridge

Pasquetto IV, Randles BM, Borgman CL (2017) On the reuse of scientific data. Data Sci J 16(8):1–9. https://doi.org/10.5334/dsj-2017-008

Pence CH (2022) Testing and discovery: responding to challenges to digital philosophy of science. Metaphilosophy 53(2–3):238–253. https://doi.org/10.1111/meta.12549

Rosenthal R (1979) The file drawer problem and tolerance for null results. Psychol Bull 86:638–641. https://doi.org/10.1037/0033-2909.86.3.638

Schmidt JC (2008) Towards a philosophy of interdisciplinarity: an attempt to provide a classification and clarification. Poiesis Praxis 5:53–69. https://doi.org/10.1007/s10202-007-0037-8

Stotz K, Griffiths PE (2008) Biohumanities: rethinking the relationship between biosciences, philosophy and history of science, and society. Q Rev Biol 83(1):37–45

Varner J (2014) Scientific outreach: toward effective public engagement with biological science. BioScience 64(4):333–340. https://doi.org/10.1093/biosci/biu021

Zollman KJ (2012) Social network structure and the achievement of consensus. Polit Philos Econ 11(1):26–44. https://doi.org/10.1177/1470594X11416766

[illegible] (200[illegible]) The [illegible] of [illegible]. [illegible] University Press, Cambridge [illegible]

[illegible]

[illegible]

[illegible]

[illegible]

[illegible]

[illegible]

Index

C. H. Pence, *Integrative Promise*, Synthese Library 499,
https://doi.org/10.1007/978-3-031-88471-9

The manufacturer's authorised representative in the EU is Springer Nature Customer Service Centre GmbH, Europaplatz 3, 69115 Heidelberg, Germany. If you have any concerns regarding our products, please contact ProductSafety@springernature.com

Printed and bound by CPI Group (UK) Ltd, Croydon, CR0 4YY
30/09/2026
02225599-0002

Palgrave Studies in Crime, Media and Culture

This series aims to publish high quality interdisciplinary scholarship for research into crime, media and culture. As images of crime, harm and punishment proliferate across new and old media there is a growing recognition that criminology needs to rethink its relations with the ascendant power of spectacle. This international book series aims to break down the often rigid and increasingly hardened boundaries of mainstream criminology, media and communication studies, and cultural studies. In a late modern world where reality TV takes viewers into cop cars and carceral spaces, game shows routinely feature shame and suffering, teenagers post 'happy slapping' videos on YouTube, both cyber bullying and 'justice for' campaigns are mainstays of social media, and insurrectionist groups compile footage of suicide bomb attacks for circulation on the Internet, it is clear that images of crime and control play a powerful role in shaping social practices. It is vital then that we become versed in the diverse ways that crime and punishment are represented in an era of global interconnectedness, not least since the very reach of global media networks is now unparalleled.

Palgrave Studies in Crime, Media and Culture emerges from a call to rethink the manner in which images are reshaping the world and criminology as a project. The mobility, malleability, banality, speed, and scale of images and their distribution demand that we engage both old and new theories and methods and pursue a refinement of concepts and tools, as well as innovative new ones, to tackle questions of crime, harm, culture, and control. Keywords like image, iconography, information flows, the counter-visual, and 'social' media, as well as the continuing relevance of the markers, signs, and inscriptions of gender, race, sexuality, and class in cultural contests mark the contours of the crime, media and culture nexus.